Other Publications by Jeffrey Warren

Stages of Life

A collection of musings and short stories originally published at Thanksgiving time in the Seattle Weekly, and currently available via free download.

Novels ...

Fifteen, Back When
and
The Adventures of Brett and Molly

This series is a nostalgic, coming of age story; set in the late 1950's and early 1960's featuring characters from a generation recognized by only a few today.

"Warren shows his amazing versatility: from the stories and musings in *Stages of Life*, to his 5 star thriller *Justifiable Homicide*, and now *Fifteen, Back When*, a touching story of 'how we were' filled with teenage angst, courage, overcoming adversity, and intertwined with friendship and first love.

"Just a great read! – **Boomer's Book Club**

AF808038

Praise for Justifiable Homicide

Synopsis

The $500 billion a year illegal drug trafficking business has a secret relationship with seemingly legitimate bankers and mainstream financial markets, which is exposed by a story in the press that reverberates on front pages around the world.

The question that even legitimate investigators seem not to be asking is: *"Where's the Money?"*

Unlikely partners Jon Reynolds, PhD (computer-encryption *wunderkind* who has a personal, vested interest in the outcome), and Katie Ruiz, a talented investigative reporter with a checkered past, devise a dangerous and clandestine plan to expose the business and the players in it — as well as to reveal to the general public the staggering proportions of the scheme. Implementing it means Jon and Katie will face cartels, violence, corruption, and an inept DEA — in addition to their own fears and demons.

This is a story grounded in actual facts that exist in the real world — our world. It is also a telling statement about the impact that ordinary citizens can have on world events — an unlikely "Power of One" tale.

Reviews

"... **a unique and truly remarkable story**." **Frank Kresen**, Editor, Kansas City

"**Justifiable Homicide is one of the most expertly executed thrillers I have read for a long time.** In my opinion, Jeffrey Warren has created an accomplished piece of writing and is well-deserving of a place alongside the top authors in this genre." **Lynne Barnes**, Stourport-on-Severn, United Kingdom, BarnseyBooksUK. Rating: 5 Stars

"**One of the top books for 2018!** Do not miss this read! Warren does a tremendous job with his characters who are well developed, believable and fun. There is a terrific balance between the story line and the private lives of his characters. The story line flows throughout the entire book with no gaps. The plot centers around "the War on Drugs" but does so from a completely different aspect - the money and asks the big question; "where is it?". Based on actual facts about the war on drugs the story is set in today's world. Warren has written the perfect David and Goliath story for the 21st century; you simply cannot help but be fascinated by the story." **Julia Walker** 5 Stars

"**This is a great book.** I thoroughly enjoyed reading it. The characters were well developed. The plot was engaging and credible. Written with so much detail that I thought it might be a true story." **Harvey – Goodreads**, 5 Stars

"**So real,** it's the world we live in, couldn't put it down." **Rachele Barger**, The Current

"**Loved the storyline!** Great job! Now in our library" **Susan Hollenbeak**, Librarian

Hundreds more five star reviews – too many to list

VIRTUAL PRIVACY

JEFFREY WARREN

Editor: Jennie Rosenblum

Design Consultant and Cover Design: the author

Printed in the United States of America

ISBN-13: 978-0-9985891-7-6

For

Being Rescued by a Rescue Dog
Thanks, Scout

Preface

From Wikipedia:

Information or data privacy refers to the evolving relationship between technology and the legal right to, or public expectation of, privacy in the collection and sharing of data about one's self. Privacy concerns exist wherever uniquely identifiable data relating to a person or persons are collected and stored, in digital form or otherwise. In some cases these concerns refer to how data are collected, stored, and associated. In other cases the issue is who is given access to information. Other issues include whether an individual has any ownership rights to data about them, and/or the right to view, verify, and challenge that information.

Various types of personal information are often associated with privacy concerns. Information plays an important role in the decision-action process, which can lead to problems in terms of privacy and availability. First, it allows people to see all the options and alternatives available. Secondly, it allows people to choose which of the options would be best for a certain situation. An information landscape consists of the information, its location in the so-called network, as well as its availability, awareness, and usability. Yet the set-up of the information landscape means that information that is available in one place may not be available somewhere else. This can lead to a privacy situation that leads to questions regarding which people have the power to access and use certain information, who should have that power, and what provisions govern it. For various reasons, individuals may object to personal information such as their religion, sexual orientation, political affiliations, or personal activities being revealed, perhaps to avoid discrimination, personal embarrassment, or damage to their professional reputations.

An example of the legal basis for the right to physical privacy is the U.S. Fourth Amendment, which guarantees "the right of the people to be secure in their persons, houses, papers, and effects, against unreasonable searches and seizures".

Part 1 – Unwanted Fame

Chapter 1

Ismael Zambada, also known as 'El Mayo', a long time Cartel boss operating in both Sinaloa and Sonora Provinces, could not stay 'retired' for long. He was not wired that way. From the safety of his hillside condo in Panama, overlooking the Pacific Ocean, he had followed via TV and the Internet the unbelievable chaos in the drug world. He was on the deck mentally going over the steps he had taken since his extraordinary cooperation with Sgt. Michael Ortiz, the previous June. He was pleased his instincts had prevailed and he had decided to support the Ortiz sting operation. He was still in shock with where that sting had led, but greatly relieved his bank accounts and other assets had escaped the seizures currently ongoing and he had been able to watch it all unfold from a safe distance. He owed Ortiz something.

As El Mayo continued to muse, one of his lieutenants stuck his head through the deck door and said, "Come in here Jefe and see what the TV is showing."

The scene showed the Plaza Melchor Ocampo, the main square of the Mexican city of Morelia, Michoacán, and the reporter was intoning the deaths of at least eight and injuries to a hundred more from a grenade attack. A group calling themselves Los Zetas was claiming responsibility.

"Oh mierda," exclaimed Zambada, "those psychopaths and butchers are active. I can see with the vacuum created by all the seizures and arrests, the lower level people are attempting to take control. It's time we return. There will be nothing left if Los Zetas expands its territory. That attack tells me they are. Gather our group in an hour, all of them, while I think this through."

Just over an hour later El Mayo held court with four subordinates seated expectantly.

"Jorge, give us the details of your purge of the Magdalena de Kino facility where we provided that Tucson cop the trailer."

"As you instructed we eliminated everyone at the staging area with any knowledge of that shipment and burnt it to the ground. All records were destroyed."

"You're certain?"

"Si, Jefe."

"Pablo, reach our Federales insider and request a contact point with Guzman for a meeting. All of you should purge from your heads any knowledge of that shipment, assuming you want to grow old and wealthy."

"Si" was repeated by the others present.

Still ensconced in their hideout in the mountains of eastern Sinaloa Province, Guzman, also known as 'El Chapo', and his lieutenants, were mulling over the message they received requesting a meeting with 'El Mayo'.

"I'm inclined to meet with him," Guzman said out loud to no one in particular, "besides learning what's on his mind; I want to know about his disappearing act before all the shit happened. We have worked together before and I see no downside if the conditions are set in our favor."

Addressing the man who brought the message, Guzman continued, "Accept the meeting request. Tell them it will be in our safe house in Mazatlán on Sunday. You will pick him up in front of the Marina El Cid office at noon. He may bring one other; no weapons."

Turning to another lieutenant, Guzman ordered, "I want the safe house area secured with our men in place. Now go."

Ismael Zambada, El Mayo, received the response from El Chapo and accepted the terms. He mapped his travel plans carefully. El Mayo was nearing sixty and had the distinction of being the only Cartel leader to have never seen the inside of a jail or prison. This fact after four decades in the business was astounding.

Zambada chose his most trusted subordinate, Carlos Rivera, to accompany him. They flew from Panama in a private King Air to a local airstrip in El Chilillo east of Mazatlán; then used a taxi to Pueblo Bonito Emerald Bay, where they stayed Saturday night. Zambada owned a condo there.

The meeting began with a strained hug between the two. Guzman opened with, "Your meeting."

"How much do you know about or remember Los Zetas as I remember you were in prison when it all started?"

El Chapo squirmed slightly at the mention of his incarceration, "Not much first hand. I saw they claimed responsibility for the grenade attack. I assumed as the brothers of our former Gulf Cartel compadre, Osiel Cárdenas, are both in custody, the Zetas are out of control. I have no use for their methods – bad for business."

"Let me refresh your memory and share my latest intel."

El Mayo went on to describe the origins of Los Zetas:

"Osiel Cárdenas Guillén, currently in custody and about to be extradited to the United States, took control of the Gulf Cartel in 1999, arguably the most powerful at that time. Turf wars erupted, challenging him. To secure his position and provide personal security, Cárdenas recruited members of the Mexican Army Special Forces. Thirty-one deserters joined Cárdenas with promises of large compensation. They were led by Guzmán Decenas, little more than a sadist. His radio call sign was Z1 and each of the other thirty was given a sequential number, hence the name.

"They began to organize kidnappings, impose taxes, collect debts, operate protection rackets, control the extortion business, and establish a sex trade and illegal immigration routes. They trafficked in various carved out zones and executed their foes with grotesque savagery. I know you remember our cooperation when we created Los Negros and all the subsequent killings. With the capture of Cárdenas, the death of Decenas, and the Cartel takeover by Cárdenas' brothers, it died down, especially after the Tampico accord was reached with the Columbians last year.

"Now with those brothers recaptured and the Gulf Cartel badly split, Los Zetas has no outside control and is on the rampage again. The current leader is Treviño Morales, another unstable psychopath. So far their activities have been confined to the east coast, Texas areas, but intel informs they intend to control the entire drug business by whatever means."

Zambada concluded with, "I propose we reestablish our former cooperation with a new partnership and prepare for Los Zetas to attack which they surely will, taking advantage of the disruption caused by those American amateurs."

El Chapo was silent for many minutes staring at some spot on an opposite wall. When he spoke it could have come from a rattlesnake.

"You come here as if everything is normal and we haven't suffered a grievous blow. And where the hell were you when the world came down around us. The fallout is still happening. And be assured those American amateurs, as you call them, will be handled and very soon. Your disappearance has all the signs of being forewarned. Other than you, wherever you were hiding, I'm the only intact Cartel left continuing business and that's because I have many layers of contingency in place. You better fucking well have a good story."

One could see the blood rise in Zambada's neck and cheeks, "You owe me a hell of a lot more respect than that. For the moment I'll excuse your words as one who has suffered great losses, is angry at everyone and everything, and is forced to hole up like a rat.

"Yes I was forewarned. My sources are extensive. How do you think I'm the only one who's never seen the inside of a cell? I moved my money to safety; I stored my inventory for a later day; I shut down my operations and decided to trust no one. You were involved with those fucking Columbians up to your ass. My sources, at present, are not available to you. If you don't want my offer of partnership and want to go it alone – so be it; this meeting is over. I'm going out for a smoke while you decide."

El Mayo stood and walked to a sliding door leading to a small deck. His subordinate, Carlos Rivera, stayed seated, his hand sliding down his pant leg to where an undetected razor knife in a plastic sheath was located.

Guzman, El Chapo, still seated and ever the pragmatist, was trying to decide what to do. Ten minutes later he signaled Rivera to ask El Mayo to come inside.

"We will drive you back to Emerald Bay in peace and safety. I will sleep on your proposal and contact you in the morning. Hasta manana compadre."

The chartered Lear 36 flight from Washington D.C. landed at the Executive Terminal in Tucson. Emerging were Dr. Jonathan Reynolds, PhD, digital signal processing and encryption expert, former Chief Software Architect at Advanced Cyber Technology in Seattle, and Katie Ruiz, investigative journalist for the *Arizona Star*. They were returning after the August White House Rose Garden ceremony, where they and the rest of their team received Medal of Freedom honors for their efforts orchestrating the greatest victory the government had ever achieved in their 'War on Drugs'.

The paparazzi are relentless. They documented the flight's take-off from Washington D.C. After a relaxing flight, both Jon and Katie thought they might finally have some peace and privacy.

That ended when they parked and walked towards the front entrance of the Arizona Star building. Three young men with cameras were positioned in various locations across the street and in the parking lot. Jon and Katie ignored them and their shutter snaps.

After reaching the newsroom floor and while heading for Katie's cubicle, her colleagues interrupted their walk. Nearly en mas they came up to congratulate them. Glancing left they noticed, for the first time ever, the office of her editor, Edgar Greeley, was dark. He must still be enjoying his notoriety in Washington, they both thought. Greeley's assistant, aka 'the battle ax' approached with a wry smile.

"Good to see you both. The boss called and will be arriving this afternoon. He asked, should I see you, please tell her to 'get back to work, there are stories to file and a newspaper to get out'." Katie just snickered and shook her head.

She went on, "Your personal mailbox is full with 220 messages. The general Star info mailbox was filled with another 200 or so for you. I sent them to archive and I have a long list of phone messages. What would you like me to do?"

Katie just shook her head again.

"Well, think about it and let me know." She turned to leave.

Katie spoke up, "Despite our differences and tests of will over the years, I can't thank you enough for your support. Please assemble all that's in your way and I'll read through it before it gets trashed. Thank you again."

The 'battle ax' named Beatrice reached out and took Katie's hand for a quick squeeze.

Jon and Katie left the Star offices late afternoon and drove to her condo on the hillside overlooking Tucson. They were reasonably certain one or more of the paparazzi followed them but at that point, they didn't care.

After opening doors and windows to clear out the stuffy air they sat on the patio looking at the view.

Katie broke the silence, "Where do we go from here?"

"At this point, no clue," Jon responded.

After Labor Day Katie was back to work and Jon was on a flight to Seattle to secure all the files and other materials prepared and provided to the DOJ and various prosecutors. He also planned to stop and winterize his Idaho cabin on the way back. They were scheduled to make a short trip to New York, reluctantly, for an interview.

Katie's first task was to organize all the emails, letters and phone messages. It was a huge undertaking as more continued to arrive. Her editor, Edgar Greeley, provided the assistance of a newly hired cub reporter named Charlene. A recent graduate of the University of Arizona journalism school, she was in awe of Katie and knew she also had been a graduate there.

Two days later they had organized the various communications into groups and categories: respond generically, or personally, or not at all.

Charlene teased Katie asking, "Are you sure all the marriage proposals are in the right category?"

Katie laughed and said, "Please start pasting my script and sending the generic emails. Use the Star Info Mailbox for those. I'll compose individual, personal responses later and send from my Mailbox. While you're doing that I'll try to catch up on today's emails."

A short while later Charlene heard Katie emit a small cry followed by, "Oh my God."

On Katie's screen she could see in all caps: YOU ALL ARE DEAD. The sender and the rest of the email were too small to read.

"Are you all right, Katie?"

Katie didn't answer, just shut down her computer, picked up her coat and left the building.

Careful to scan her surroundings upon departing the Star offices, on the drive home, and in both her garage and condo, she was slightly shaking the entire time. She fixed herself a stiff margarita and sat to collect her thoughts before calling Jon.

Just after 5 PM in Seattle, she called. He answered on the second ring and could immediately tell from the tone of the greeting something was wrong. He told her he was with Karl and Marie.

"Jon, love of my life, I'm not sure whether I'm more afraid or more pissed. Please just listen. I received an email this morning in the Star Info mailbox from someone with an address of sinaloa666 at Hotmail.com with a subject of Greetings. The first line was YOU ALL ARE DEAD. Below was added, 'you picked the wrong people to try and fuck over – you won't know from where it comes or when it comes – enjoy your short time left living in fear' – there was nothing else. I don't know how to do a trace route for the email. Do you?"

When Jon didn't immediately respond, Katie asked, "Are you there?"

"Yes. A trace will only lead to Microsoft. You would need their information on that account. It could conceivably be done with a subpoena. You might try contacting Wes. Before we go any further you will need to warn Michael and Edgar. I wish I was there to hold you tight."

"Me, too."

"Do you feel safe?"

"I think so. Everything is locked tight and no sign of anything suspicious here. This really sucks."

"To say the least – if you're OK let me brief Karl and Marie and have a chance to think. Talk soon. I love you."

"Love you too, bye for now."

Both Karl and Marie's facial expressions showed a combination of shock and fear as Jon finished relating Katie's information.

Marie spoke first, "Do you believe the threat is real? How can we protect each other?"

"Yes I do and to your other question: I don't know right now. That's the worst part."

Chapter 2

"Good evening, everyone. Tonight we are pleased to welcome back Kate Ruiz, investigative reporter for the *Arizona Star* in Tucson, author of two remarkable pieces of journalism that nearly everyone from around the world who follows the news has seen or read. And also joining us is Dr. Jonathon Reynolds, who needs no other introduction. I'm Barbara Walters, and this is a special edition of *20/20*.

"Thank you for joining us tonight and congratulations on your well-deserved receipt of the Medal of Freedom. The fallout from your efforts and stories just continues. The latest figures from the DOJ claim over four billion dollars of illicit drug monies have been seized and all Cartel bosses known to still be alive are in hiding, including El Chapo.

"Would you please open tonight's program by describing the other members of your team, as you call them."

"Jon you go first," suggested Katie.

"Well, Karl Oberling is my long time employer at Advanced Cyber Technology and together with his wife Marie are very best friends. They were instrumental in all the database processing that traced the drug payments. The super cops in Seattle, Officers Thomas Murphy and Don Minelli plus their Lt. Fisher, not only protected us from the DEA operative who killed their bent agent and came after us, but were instrumental in tracking and interdicting the actual drug shipments. Sgt. Michael Ortiz of the Tucson Police Department I'll leave to Katie."

On cue she said, "I've known Michael for many years and his talent, instincts, knowledge, and integrity astound me. He set up and orchestrated the sting operations that provided the path leading to all the drug money transfers. We could not have done this without the great support from my editor at the *Star*, Edgar Greely and the Pima County DA, Wes Davis. In addition we had two young grad students who wish to remain anonymous, helping put together the massive amount of data the various prosecutors are using today. All in all, a truly amazing group who deserved their medals and more."

Walters looked directly at Jon, "A number of media articles have suggested that you, with your genius level knowledge and background in encryption technology, essentially hacked the banking money transfer networks. Care to respond?"

"I'll let the US Attorney General's investigation respond. Next question, please," Jon said evenly.

"Alright, now tell us what's next for you, both personally and professionally."

Somewhat sheepishly Katie answered, "Jon proposed and I accepted. When hasn't been decided." Jon just grinned.

"Congratulations to both of you," Barbara smiled broadly, "and professionally? How does one even attempt to replicate your amazing journalism?"

"My editor, Edgar Greely suggested I write a weekly column and associated blog. He is working on syndication. We'll just have to see."

"I will look forward to reading and following you, and for you, Dr. Reynolds?"

With a suppressed laugh, Jon answered, "At this point I'm clueless. After all these years and the recent intensity, I'm looking forward to a break and being a good husband and house-husband. You can probably guess I'm very uncomfortable with all this notoriety and don't know right now how to slip back into our desired and needed private lives. That's about as straight as I can be."

Walters paused for a moment before continuing, "I can almost feel your anxiety. Thank you for sharing and your honest feelings. Thanks to Katie and Jon and we'll be right back with our next guest."

Zambada and his second, Carlos Rivera, were picked up in a non-descript sedan at 9 AM from Emerald Bay. They were driven east out of the populated area of Mazatlán. At a Pemex gas station they were escorted to an outwardly decrepit 28 foot Chevrolet panel truck. Inside the truck's box sat El Chapo and two others. The inside bordered on opulent: soft leather swivel chairs, a bar with a sink, refrigerator and ice maker. There was a corner communications desk with a large monitor and an air conditioner hummed in the background.

El Mayo and Rivera stood on the ascending lift gate and entered the truck's box.

Guzman, El Chapo, waved a hand, "Welcome and be seated. We have a several hour drive and this is my preferred method of transportation. Help yourself to the bar; it's always open."

Zambada and Rivera helped themselves to very strong coffee, took a seat, and sipped as the truck got underway. They glanced around the interior and waited for Guzman to break the silence.

After a few minutes the rough road bumping decreased and the ride smoothed on pavement. Addressing the others in the truck, he went on, "I have decided it would be in both our best interests to cooperate. A partnership as you proposed presumes a level of sharing that I'm not willing to do at present – that may happen in time. As a demonstration of my good faith, today I will show you something that will allow your inventory to begin moving north again with no worries. We'll be there in about an hour so please enjoy the ride."

Zambada simply nodded and said nothing.

"Tell me El Mayo how did you know about what came down on us?"

"I have many sources in Sonora and I had used Abram's Trucking for many shipments. I learned they had been compromised," was all Zambada would say.

"While we all used the readily available trucking networks, I decided another cross border path was needed, one that I learned about from those Tijuana assholes. You will see it shortly."

"How did you escape capture and how bad did the bank account seizures impact you?"

Guzman shrugged, "We lost about sixty percent of our balances, but we're not hurting. The amount of money is almost obscene and managing it is actually more difficult than selling our products across the border."

A short while later they arrived in the small Mexican town of Agua Prieta, across the border from another small town named Douglas, Arizona. They pulled into a moderate size metal building with no windows and a single roll up door, opened with a remote. It was dark inside and the driver got out, went to a known vertical post and flipped the light switch. Then he opened the panel truck's rear door and everyone disembarked using the tail gate lift.

"Follow me El Mayo; I wish to show you something."

The driver appeared with a six foot long steel rod with a hook on the end. He inserted the hook into a ring on the floor and lifted a hinged door. Two other Guzman men fully opened the heavy trap door and a lighted stairway could be seen.

El Chapo was beaming, "Down those steps lays a 250 meter long tunnel that comes out in my animal feed store business across the border in Douglas. I have five other such tunnels."

"Increíble," said Zambada with total admiration and took a few steps down.

On the ride back from Agua Prieta, Guzman and Zambada discussed the details of their budding partnership. Zambada would reopen several of his facilities; they would share Guzman's transportation to and through the tunnels. Guzman was very pleased to have access to Zambada's inventory as his was painfully low because of all the prosecutions and seizures – especially cocaine. He had been forced to source from Peru and Bolivia. Further complicating matters was the Mexican government's decision to ban the import of ephedrine from China and India. El Chapo's meth super-labs were nearly shut down.

For use of his transportation system, Guzman demanded twenty percent. Zambada, with no real choice, agreed but stated there would be no commingling of their finances. Guzman said, "Of course not. It will take both of us a few months to recover and regroup. We will need to communicate. We use satellite phones; what about you?"

"The same."

They exchanged numbers and agreed to meet again in the New Year. Zambada and his lieutenant were dropped off at Emerald Bay.

Chapter 3

Katie was fussing around in the kitchen but the sliding glass door was open to where Jon was stretched out on the patio chaise lounge.

"Regarding domestic affairs, I'm used to you taking charge, issuing orders, and generally being a highly seductive alpha woman. We've been marking time for almost three months. What gives, sweetheart?" Jon asked.

It was the day after Thanksgiving. She came out with her cup of coffee and the pot to refill his.

"Don't put this on my doorstep, big boy or should I say clueless boy. Remember your words. It was you who said we should keep a very low profile while governments and prosecutors seize the money and pursue their prosecutions and more importantly be vigilant to protect ourselves. Even with help it took well over a month to sift through my messages, eliminate the ridiculous and obscene, answer those I chose and still it continues. Do you know how many marriage proposals I've received? Don't go there. The calls and emails are getting in the way of doing my job. There isn't a week that goes by without some paparazzi photo and associated BS story about us on the Internet. At least the mainstream media has been somewhat respectful."

"It's time we had a serious, personal discussion and now seems as good a time as any."

"As for marking time, that is inside your introspective head. Let's see: in September, when the threat email arrived, you were in Seattle for ten days to secure all our files; then you spent over a week in Idaho winterizing your cabin; and after our quick trip to New York you've been hunkered down here. Outside of that I keep trying to write my weekly piece while you withdraw into your own thoughts. I've been very patient – not my nature – but that's about to run out. Where'd your comment come from? What'd you do, decide to wake up?"

Silence lingered. Katie sipped her coffee and waited.

"Boy have I screwed up. Old habits, I guess. I'm really sorry, please forgive me. I've been wrestling with what the future holds and I can't seem to come up with a clear path."

"You're not off the hook so easily. I've told you before that knowing the outcome is highly overrated if not impossible. And don't pull your analytical crap on me. Doesn't it seem reasonable the future, our future, should be mapped out in a mutual discussion? Or should we just write off our whirlwind relationship as project based and sex driven? Is your proposal on hold?"

More silence.

"I am just consumed by fear. We can't ever again let things get this out of control. I really am truly sorry. I hope you know I love you and you're my soul mate."

Katie leaned over and kissed Jon tenderly. "I love you too, jerko."

Tension left the air.

"Are you happy?" Jon asked.

With a moment's pause, Katie responded, "That's the damndest question you've ever asked me. My immediate response is – with us absolutely – with all the winds swirling around us, absolutely not."

"My sentiments exactly. We need a course correction. We deserve to be happy, at peace, and have personal privacy."

"Do you really believe that's possible?"

The next morning, Saturday, Jon was frying bacon and preparing batter for waffles. Katie was trying to pull the covers over her head and go back to sleep or at least snooze awhile longer but the smell of frying bacon prevented that. She donned her robe, staggered out of the bedroom, and accepted the cup of coffee Jon held out as she passed by on her way to the deck. Jon knew better than to say anything.

"Breakfast is served," he said awhile later.

"I'm in recovery mode lover, please bring it out here."

They sat on chaise lounge chairs with a glass top table between them and devoured breakfast. Katie ate her waffle faster than Jon for a change.

"More coffee," she muttered holding out the cup.

It had been a wild and unusual Friday night for them. El Matador, the restaurant owned and operated by Katie's relatives, had thrown an amazing party, in part to honor their revered 'daughter' but also to thank their regular and loyal customers. Michael and Corinna Ortiz were there, as was Edgar Greeley and Wes Davis, the Pima County DA, and his wife. The sangria hit Katie just after midnight. She should have stayed with the margaritas. Jon carried her home and into bed shortly before 1 AM.

"My head hurts," she continued with a question, "Besides you, who else needs an apology? I haven't been out-of-control drunk for over fifteen years. What the hell happened?"

Jon reached over and took her hand. "You owe no one an apology. You were the perfect lady all night, albeit very funny. Guess it was beyond time to let it all out. Relax and recover. Want some aspirin?"

Katie grinned sheepishly, "No, I'm returning to the living. Thanks."

She continued, "The last thing I remember is dancing with Hector. How embarrassing. I can't believe the whole thing. Will I see myself on the cover of some Tabloid?"

"Relax and I'll get you more coffee."

It was a cool day. The temperature not expected to exceed 70. By midafternoon Katie, still in her robe, was almost fully recovered although now stretched out on a deck chaise lounge. Jon took a chair beside her.

He said, "How about we pick up the 'future' conversation of yesterday? I've come to some conclusions and have some ideas to share and discuss. Clear headed enough?"

"Actually I am. Was just laying here wondering if some gunman will suddenly appear and put a bullet in us. We're really at a crossroads in our respective lives. I'm also certain when we embarked on our project – no mission is a better word – neither of us thought about what the future held then."

"At the risk of invoking the analytical process, please listen and respond as I throw out thoughts and ideas. This will delve pretty deep. You know it's really cool to have someone you love, can talk with, and trust will understand."

Katie just smiled.

"Let's begin by summarizing our current situation. As a direct consequence of our research, its risks and your stories, we became known worldwide. We also became enemies of those drug cartel survivors who we ruined and/or exposed. To the celebrity obsessed media we became famous, in no small part because of your beauty. We have maintained a small circle of friends and a slightly wider circle of good acquaintances. We have earned the respect of various government, police and military personnel. All the foregoing have resumed somewhat normal lives. Not us."

Katie was alert and focused, her attention on Jon's narrative.

"We have spent over three months dealing with the arrows coming our direction – well-meaning or otherwise, looking over our shoulders for any sign of enemy attack, pretending to lead normal day-to-day lives, using each other as protective buffers against our real fears and the outside world. Yes, there were a few breaks in the routine, but the reality was never far away.

"Enough is enough. I can't live this way. On the plane ride here from Washington DC, I asked you to marry me. No sooner had we arrived than reality struck and how in God's green earth could I offer you what you deserve with all this unsettled baggage and then the DEAD email came in. I have a plan or better put a suggestion, but first, to use your favorite expression, your turn."

"Pretty damn accurate Mr. Analysis. As this is a gut level conversation, let me add my few cents. Prior to you I was the most hard-nosed, self-willed woman you can imagine. I fought for and maintained my independence. But love changes that. One wants to give and reach mutual understanding and bliss – you know, happily ever after. Sure, my natural tendency to want to control and direct is still there, but I'm really quite happy and content to defer to you when it comes to major and potentially life threatening matters. That's been proven. Short form: I do not want or seek to be a celebrity; I need to be recognized for my abilities and I want to be your wife, partner, lover, and very best friend. Please give me your view of our future."

Jon hesitated for a moment looking intently at Katie, thinking what a marvel; I don't have adequate words to describe her; I know I just love her unconditionally.

"Sweetheart, yesterday I used the phrase 'course correction'. I don't know where that came from but I've been thinking about it a lot. Let's progress logically through the issues – first, where to live. While I love the Idaho cabin, we can't live there – it's a place to visit and enjoy. Your Tucson condo here is not someplace I can live year round and it's too well known to our enemies; it's another good place to visit our friends here. Seattle has no interest for me for a variety of reasons. We need a home base; more on that later.

"Second issue is our employment. You are a natural investigative journalist. I should think it's pretty obvious you are well beyond the Arizona Star crime beat and I can't see you forsaking journalism, even for me. Your weekly column has been well received and the syndication that Edgar arranged is now what, 171 papers? As for me, I haven't decided what I want to do when I grow up but rest assured I can do it from any geographical location, so can you. With our so-called consulting fee stashed safely offshore and my investment accounts from the ACT stock, we don't have any money issues.

"Third issue is our safety. I'm still working on that one. Please give all these thoughts your consideration. We'll talk again in a few days, but some decisions need to be made soon."

The following Tuesday, just before Katie was heading out to work, Jon's cell phone rang. The caller ID showed Karl Oberling. Jon answered and before a greeting could be offered, Karl said, "Bring up the Seattle Times website."

Jon walked to Katie's desk and computer to comply. The headline blared across his screen in all caps:

SEATTLE POLICE ASSASSINATED
MEXICAN DRUG CARTELS RESPONSIBLE

"Oh my God," Jon whispered into the phone.

"Read the coverage and call me back," Karl said and disconnected.

"Katie," Jon yelled, "com'ere quick."

Together they read all the newspaper articles including a poorly written editorial. Before calling Karl back Jon made sure all their doors were locked, scanned the visible area around Katie's condo while she called Michael Ortiz and her editor Edgar Greeley.

The online newspaper edition reported that Homicide Officers Thomas Murphy and Don Minelli were seated in a downtown Seattle, 5th Avenue sandwich shop, when a single hooded gunman calmly walked up to their table and shot both in the forehead. The gunman walked out the shop's front door and disappeared. Statements are being taken from the restaurant customers and Seattle Police are requesting any information from the public who may have witnessed the shooting or the shooter leaving. The restaurant did not have video surveillance.

The sidebar articles described the officer's involvement and role in the recent multi-national drug cartel busts – currently ongoing. Without any speculation, the officer's murders were blamed on Mexican Drug Cartel revenge.

Karl answered before the second ring. He began to speak immediately.

"Who's next? That's Marie's and my biggest worry. The most insane thing is we found out about this from TV Breaking News, not anyone else and it happened yesterday at noon. Lt. Fischer is somewhere in hiding, but we now have a patrol car across our street. Are you safe?"

Katie spoke first, "You're on speaker Karl and thank God you and Marie are safe – at least for now. I just called Edgar to warn him and left a message for Michael."

Jon's mind stopped spinning as he began to focus. "Karl, please do as I say. Get packed and ready to leave with Marie and Stephanie. I'll have a plane standing by at the King County airport executive terminal. Only I will know where it is going. Have that patrol car take you there after I call back with the timing. Are you good with that?"

"Will do, friend."

Katie added, "I need to track Michael down. I have a bad feeling. Bye for now."

Katie tried Michael's cell phone again and when it went to voice mail, called the Captain at police headquarters.

"Thanks Ms. Ruiz for the heads up and your concern. Ortiz is part of an arrest operation at present, and as he's surrounded by others, I'm sure he's safe. I can't tell you more but will follow up."

Through an informant a warehouse had been identified to be a holding area for illegal immigrants and their smugglers. Michael Ortiz and his officers, accompanied by a SWAT team, stormed the warehouse and secured the area. In the process four smugglers were arrested and eight illegal Mexicans taken into custody as drugs were found on each.

Ortiz instructed one of his officers, a young patrolman named Cody, to drive the panel truck they had seized to the impound yard. Ortiz would follow to bring him back to HQ.

Cody stopped in front of the chain-link fence, got out to unlock the gate but didn't know the combination for the keypad. Ortiz got out of his car and as he walked towards the gate, his peripheral vision caught someone to his left. As he turned his head, a gun discharged and a bullet hit his chest. He staggered and another shot was fired. Ortiz went down.

Operating purely on instinct and good training, Cody pulled his Sig and put a bullet through the shooters' left ear. Ortiz was moving when Cody rushed up to him.

Ortiz looked up and said, "Thank you God for Kevlar, but damn it hurts."

Chapter 4

Michael was in the ER getting his chest taped when Jon, Katie and Corinna walked down the hallway. The Captain was seated on a bench outside with two other officers standing alongside. They looked in and Michael raised a hand signaling 'just a minute.'

Katie began to speak when the Captain held up his hand, "I was wrong and over confident. I'm very sorry."

Michael came out fifteen minutes later and gently gave his wife a hug and peck on the cheek. Tears ran down Corinna's cheek over the kiss.

The Captain stated his requirements, "We will drive you home. I want you to stay there and rest. You can dictate a statement over the phone to a transcriber. Your home will be covered 24/7 until we can sort this out. Until then you are on paid, administrative leave. I'm so relieved you are OK, Sgt."

Jon and Katie followed the police cars to Michael's home and waited while the police entered and announced the house was secure. Two patrol cars remained, one in the driveway and the other across the street in front. The officer in the driveway circled the house every ten to fifteen minutes.

Inside, Jon, Katie, Michael, and Corinna talked through the entire situation without reaching any conclusions. The Captain had assured them Edgar Greeley would be covered. The identity of the shooter had not yet been determined.

Jon booked a Lear 36 charter from Seattle to Tucson with a potential continuation for 4 PM PST that Tuesday. He communicated the departure with Karl and said he and Katie would meet them at the airport upon their landing.

They returned to the condo and Jon was deep in thought on the deck while Katie fixed sandwiches for a late lunch.

Upon joining him and taking a bite Katie asked, "Care to share your thoughts?"

"Everyone's safety is of paramount importance but I'll be god damned if we will go into hiding and spend years or more looking over our shoulders in fear. The only logical course of action is to take this to the enemy and put an end to it. I have a couple of ideas how to do it."

"For your entire public persona as the calm intellectual, underneath you are one scary dude. Lay it on me. I agree with you and am onboard."

The Lear Jet charter taxied up to the Tucson executive terminal shortly before 9 PM local time. Karl, Marie and their fifteen year old daughter Stephanie deplaned and were greeted by Jon and Katie before reaching the terminal. Jon instructed the efficient staff to load their baggage into the waiting Lincoln limo and the pilots to take the night off awaiting further instruction before noon the next day. A private security team, recommended by Michael Ortiz and hired by Jon, formed a perimeter around the terminal.

The entire group departed. Jon with Katie driving her car, the limo with the Oberling's following and the security team shadowing, made their way to Katie's condo. The limo left; the security team deployed outside, and Katie announced, "Marie, you and Karl have the guest bedroom over there. Stephanie you can bunk up on the couch or outside on the deck – your choice. Your bathroom is there."

The group was standing at the railing on Katie's deck, enjoying the warm late November evening and the Tucson city view below when Marie said, "It's so beautiful, peaceful and inviting down there, but we know that's not the truth."

Jon squeezed her shoulder and responded, "It will be again. Let's talk in the morning. Sleep well. We all need that."

Whether from the worry and stress, the sound of the air conditioner and coffee maker or the early daylight, everyone was awake and alert the next morning before 8 AM. It was decision time and everyone instinctively knew it, even Stephanie. It was also internally acknowledged that each of the group looked to Jon for insight and direction.

After morning greetings and pleasantries Jon called Michael Ortiz.

"Are you able to join us here and bring Corinna? ….. Great, see you in an hour."

Katie's neighbors were home so the deck door was closed. The assembled group had taken whatever haphazard seating was available. Jon sat on the small hearth of a seldom used gas fireplace. After several minutes of uncomfortable silence, it was Katie, seated on a large pillow leaning against the deck door, who began.

"Let's talk about shared responsibility first. Our lives are at risk; team members have been killed and attacked; we're all uncertain and frankly scared shitless. Whose fault? I wrote the stories, Jon provided the analysis, each of you contributed to the data. We're in this together and we will come out of this whole, intact and together. Get that straight."

Michael Ortiz could hardly suppress a grin; Corinna squeezed his hand; Stephanie looked to her parents; Karl put his arm around Marie and said, "Damn straight."

Jon didn't take his cue for over a minute.

"I love you all; simply an amazing group of friends. It's humbling to have your friendship, support, and strength at this point. At Katie's urging, allow me to summarize what I see and then we can discuss.

"I would like to open this conversation with something I told Katie a few days ago: our mutual safety is of paramount importance but I'll be god damned if we will go into hiding and spend years or more looking over our shoulders in fear. The only logical course of action is to take this to the enemy and put an end to it."

"Do you really think that's possible?" asked Corinna.

"Yes I do."

"How?" asked Michael.

"What do we know for certain? From within the Mexican drug business our collective deaths have been ordered. There is no shortage of killers available to carry out the order. We're not capable of identifying those killers. There is no way to hide or have sufficient protection around us 24/7 to prevent an attack. The only way we can become safe and lead normal private lives is for the order to be rescinded, canceled.

"Next question is who gave the order, or better asked who is capable of canceling it? And what motivation could be offered for such cancellation along with sufficient guarantees for our safety?

"We need to answer those two questions."

Michael responded first, "I don't dispute anything you said, Jon, but I'm not seeing your method. We're not going to do house to house searches in Mexico."

Karl spoke next, "I'm thinking the person who gave the order is one of those Cartel bosses not captured and number one on that list is that El Chapo character."

"Exactly," Jon continued, "and whether it's him or not, they have to be able to communicate with each other and they're sure not at the same physical location. What was the device those attackers at my cabin carried and we used to locate them – a satellite phone?

"Michael we need to know what is happening with the investigation into you being shot. Did the shooter have a phone? What other personal effects did he have? Did they find a car or figure out his transportation? That's first. Second is to choose a temporary safe location for all of us, including a base of operations. I believe trying to resume normal lives right now is out of the question. Any thoughts on that?"

Marie asked, "Do you really believe we're all equally at risk? I mean they attacked only the police so far. Dropping or hiding out, even for a short while disrupts so many things." She looked over at her daughter.

Katie answered, "We have no way of knowing for sure. We can either assume the worst or take our chances while being vigilant. I suggest we bunk up here in Tucson for at least a few days while we pursue Jon's idea. We can make a final decision by Sunday."

"I like that idea," Marie replied.

"Let me see what I can find out," Michael added.

Karl, Marie, and Stephanie took Jon's truck, with an escort, to the Saguaro National Park. Katie drove, also escorted, to her *Star* office. Jon's phone rang. It was Michael.

"There are days when I think our department has their heads stuck so far up their ass they're in danger of disappearing."

"What happened?"

"I called the Captain to inquire about the investigation. He told me they assigned it to … you don't want to know. The long and short is there is no current investigation. This responsible idiot claims his caseload is so large he won't be able to devote significant time for a week or two and besides it's probable some Mexican we can't ID or connect to anyone else was the shooter and it's probably connected to some of my past cases.

"I called Cody, the patrolman who basically saved me, and asked for his help. He called a few minutes ago. There was nothing on the shooter – no ID. His prints are not in any US database. It seems they stole a rental car on the next block over from the impound yard. Inside was a duffel bag with clothes and guess what, a satellite phone. The dumb shit with the investigation hasn't connected the dots. I'm going to the car rental agency now and collect the contents. Call you as soon as I know something."

Jon had nothing else to do so he decided to reconcile his various bank accounts, something he had put off for over four months. He had spent a lot of personal money in support of the research leading to Katie's stories. An hour later he had created a spreadsheet summarizing all expenditures.

He had spent almost $400,000 from his personal accounts for everything from marijuana purchases to airplane tickets and charters. The current balance in his Piper Jaffrey account was $51,434,370, primarily from his Advanced Cyber Technology (ACT) stock. He had $1 in Delta International Bank, $1 in State Street Bank and Trust, the company he and Katie had destroyed for laundering drug proceeds, $35,000,000 at Credit Suisse in the Bahamas seized from Ismael Jimenez-Sanchez, the Columbian drug boss, and $74,542,660 in Zurich seized from Theodore Devlin, the mastermind of the drug money laundering and investment scheme. He thought that was a hell of a lot of money. He was unique in that he had never worried about money since ACT went public. His current belief was money only bought one thing and that was freedom. He would spend freely to achieve that result.

The buzzer on the front door of Katie's condo rang. Jon answered as he was confident the outside security was in place. It was Michael Ortiz.

"I couldn't convince our dumb shit department to send the lab to go over the car, but here are the contents." Michael placed the contents of a plastic bag plus a duffel bag on Katie's counter.

Jon's focus went right to the SAT phone.

It was a Globalstar phone similar to what Jon had recovered during the attack on his Idaho cabin. He turned it on and was connected to the network in less than a minute. That told him the account behind the phone was active. He did not try any other buttons because he was unfamiliar with its operation. He sat down to think; Michael waited.

Jon's thoughts were interrupted by Katie's call.

"Do you want the good news or the bad?"

"Always the bad first."

"Time Magazine has us on their cover. We're numbers one and two on their Most Influential People List. It's on newsstands today. I received a congratulations call and request for comment from ABC. The magazine story has our bios and a complete overview of the stories and background. I feel blindsided and don't understand how they could publish without our knowledge – probably some public domain crap. The only team members not mentioned were the grad students."

"Holy shit and what could possibly be good news?"

"We're having dinner at El Matador. Hector is preparing a feast in the private dining room. I called Karl and Marie on their cell phone. You and Michael and Corinna get there, OK? You might want to have your security folks standing by. We'll get through this. See you love." She hung up.

Part 2 – Counter Attack

Chapter 5

On Thursday morning they were gathered again in Katie's condo. Several copies of Time Magazine were lying around. Jon could not understand why Katie didn't seem as upset as he was.

She said, "C'mon Jon, don't withdraw from us. You must know our wish for anonymity was a pipe dream. Besides it's a great picture and nothing in the article hadn't been published or said before. I don't think it changes anything."

Marie added, "She's right, you know."

"Alright," Jon said, "let's focus. I was up late last night researching satellite phones and this brand in particular. I feel comfortable examining it.

"Karl, what I learned last night took me back to one of the first projects I worked on at ACT. Do you remember that small contract you got from Hughes Aerospace for that encryption algorithm?"

"Of course I also … holy moly, Hughes was the manufacturer of all those communication satellites. You wrote that code. Do you think it's still in use today?"

"We're going to find out. I want you on the Lear as soon as possible. Find the files. They should still be in the ACT server archive. Bring them back here. OK?"

"OK."

The conversation had everyone's rapt attention. Jon went on to explain.

"Globalstar, the company and network for this phone operates about forty low earth orbiting satellites that send and receive calls and data between them and a number of earth stations. Those stations interconnect to both the landline and cellular telephone networks. The GMSS or Global Mobile Satellite System manages various providers. These phones are only identifiable by a unique country code prefix. Globalstar has been assigned +881 plus the numbers 8 or 9. They allocate a somewhat random local number to their subscribers in the country they're based. The phone numbers are not traceable; however most devices include a GPS unit. That's how the NSA determined locations from the two phones of our attackers. All voice and data communications are encrypted. The specific algorithm is proprietary to the network. As such, calls cannot be overheard. Unless there are two satellite phones connected to each other on the same network, locating them is virtually impossible.

"Until I can get a look and refresher from that old ACT contract, we're on hold."

Marie asked, "Can Steph and I catch a ride with Karl? We're thinking Stephanie can stay with my sister, continue school and will be safe there. I plan to come back and help in any way."

"Of course," Jon answered, "before you go let's turn this phone on and press MENU."

He did so and noticed the Not Connected icon stayed lit. He said out loud, "Yesterday this phone connected within a minute. Now it isn't connecting. I wonder if it was removed or shut down. Maybe the person behind all this realized we probably have the phone of his shooter."

Jon pressed the MENU button again and among the icons one was RECENT. He pressed that and a list of nine calls with numbers appeared. Three of the numbers had a 206 area code – Seattle. Another was 520 – Tucson. The other calls were to the same number – the location was unknown. The displayed number was 88189165240.

"Michael, do you know where in Mexico the prefix 916 might be used? Or can you find out?"

Michael made a call, then another, made his request and was placed on hold. He then listened and hung up, "There are several, not an easy lookup. My source will email the results in a short while."

Karl suggested, "Get your plane ready while we pack up a few things. Do you want us to just turn around or would tomorrow morning be OK. We would like to get Stephanie settled."

"Tomorrow is OK but you better be covered all the time and we don't know where Lt. Fischer is."

"I can take care of that," Michael replied.

Guzman was again in his truck traveling to one of his distribution facilities. A phone rang at the truck's communication desk and a Guzman subordinate left his chair to answer it. He listened, nodded several times, said OK and hung up.

Guzman's subordinate took his chair and said, "We lost Javier."

"How? Where?"

"In Tucson, he was shot and killed. The cop survived."

Guzman turned to his lieutenant and said, "Tell me about Javier."

"You know he went to Seattle first about those cops. Their Lt. disappeared immediately. He was there for a week and unable to stake out the Oberling's because of where they live; also they rarely came out. The man seemed to be working from home and when he did go to his office it was inside a large SUV from his garage to the private parking in his office building – no opportunity even over the holiday weekend.

"You also know our Seattle distributor is in jail awaiting trial. Bail is not possible. We have had to deal with a very unpleasant and demanding gang leader. Javier decided to go to Tucson and flew there last Monday night.

"According to our sources there, Ortiz was wearing a Kevlar vest at the time and only received a cracked rib. Javier was shot by another cop with Ortiz. What do you want me to do, Jefe?"

"Who is our next best with perfect English? I want them all D E A D."

"I will find out who is available and let you know."

Jon drove, a security man followed, Katie was dropped off at the Star and the Oberling's taken to the airport. Jon gave Stephanie a big hug and whispered, "Everything will be alright." She gripped him tightly.

To Karl's pleasant surprise a SPD squad car was parked at the door of the Executive Terminal at King County Airport when the Lear landed. Michael had come through. An overnight rental car was provided and with the squad car following they first went to their home and packed what Stephanie would need. Ten minutes later Marie's sister pulled into the driveway. Karl and Marie said emotional goodbyes to Stephanie, then headed for the ACT offices.

The offices were dark and locked as it was after 9 PM. Nevertheless the SPD officers cleared the space after Karl unlocked the main doors. Marie made coffee and chatted with the officers in the reception area.

It took Karl over an hour to find the old files, copy those and the source code Jon had written. As his access to the servers would be noted, he left a brief note on the CEO's desk. They were home before midnight and fell fast asleep. The officers told them regular patrols would be in the area and to call them when they left in the morning. They offered their phone number.

The Lear was in flight and Jon was headed to the Star to wait for Katie when his cell phone rang. It was Michael.

"Just received the email; the 916 prefix is used in three different locations under different area codes. Two make no sense in our circumstances but one does: Mazatlán."

Jon said thanks and accelerated slightly as his mind began to churn through the possible next steps. He called Katie's cell and told her he was in the parking lot and would wait for her. He declined her invite to come up to the newsroom claiming he was in thinking mode. She knew what that meant.

She came out forty-five minutes later and on the drive home told Jon about a staff meeting where the current state of the cross border drug business was discussed. The shipments have continued almost unabated. One large shipment destined for Los Angeles had been intercepted and several smaller ones had been found in the Phoenix and Albuquerque areas. Authorities had no information how these had crossed the border and were blaming everything from corrupt customs inspectors to NAFTA in general. Both were probably true she commented.

Katie went on to describe the recent rise in killings as Cartel underlings were vying for control. Many of their former bosses were in jail or back in prison. In particular a group from the Mexican east coast calling themselves Los Zetas was being accused of brutality against both rivals and civilians.

Jon listened and just shook his head.

Upon their arrival at the condo Katie took Jon's hand and pulled him into the bedroom. "It's been awhile; we're finally alone; I know how to relieve your tension."

Making love started slow but soon revved up to an intense climax. They rolled onto their backs holding hands and were soon asleep.

Friday morning dawned with a hint of unusual rain. Jon was informed the Lear had taken off on its return trip. Karl reported success with the old files.

Jon looked at his watch estimating the touch down time and asked Katie, "I'll go pick up Karl and Marie. We need a safe base of operations. Would you research available real estate in and around Mazatlán?"

"Have you lost your mind?"

"I don't think so. What would you consider to be the safest place in northwest Mexico?"

Without a moment's hesitation she answered Cabo San Lucas, tip of Baja, directly west of Mazatlán across the Sea of Cortez.

"Check it out please. I've got to go and could you contact Wes and try to subpoena phone records for those 206 and 520 area code numbers. I wrote them down here. Also get his opinion of a subpoena for the Hotmail account records on that email."

"I don't mind doing research or running errands. Just remember we are a team. You haven't shared one damn thing with me all week. You've been withdrawn and I don't like it. Get a clue Jon. You don't have to provide all the answers."

"Very, very sorry – we are a team and you have been very patient and supportive. I only know one way to think and analyze."

"Apology accepted." They shared a long embrace.

An hour later Jon returned with Karl and Marie.

They snacked on cheese, apples and crackers for lunch when Jon asked, "Katie did you get anything? Any news from Wes?"

"A number of possibilities but I really don't understand what you're up to. Wes will let me know ASAP."

"I have a potential plan but first need to examine the files Karl brought back. How about you let us do that. Why don't we bring in some pizza for dinner? Would you ask Michael and Corinna to join us? Hopefully my plan will take shape and I can explain then. Please indulge me."

Katie shrugged and nodded in the affirmative.

Jon and Karl huddled over the dining table, their materials spread out. Katie and Marie stripped beds and did laundry. Shortly after 5 PM Karl and Jon grabbed a beer and went out to the deck as the threat of rain had passed.

"Was the Pentagon work you and Brad did going on at the same time as the Hughes contract?"

"For certain and our technologies were used in parallel. The Pentagon was focused on decryption and Hughes on encryption. Under the various proprietary and non-disclosure agreements the left hand didn't know what the right was doing – typical. You have that familiar gleam in your eye, friend. I'm beginning to see where you, or should I say we, are going."

"Let's discuss with the whole group."

The pizza arrived at the same time as Michael and Corinna. More beer came out and they devoured three large supremes. The six of them were seated awaiting Jon to open the conversation.

"A few months after Karl hired me at ACT, Hughes Aerospace requested a preliminary design and bid to develop an encryption algorithm for them. At the time ACT's primary line of business was developing decryption software for the Pentagon and by connection NSA. This was before ACT got into encryption methods for banks and their networks that you know all about. I was assigned by Karl to handle the Hughes' request.

"I flew to the Hughes facility in California to present the preliminary design, became very familiar with their requirements and area of application, then returned and completed the contract by providing a proprietary encryption algorithm to be used in voice and data communications for their satellites and network."

The group now had an ah-ha look on their faces.

Jon went on to explain that Hughes was one of the world leaders in the manufacture of communication satellites and the systems used for voice and data transfer. Hughes has provided all the satellites for the Globalstar network including the encryption algorithms used for voice and data transfer.

Karl took a nod from Jon and continued, "We will purchase our own satellite phone from Globalstar and once activated record a call. From that recording we will examine the packets and attempt to decrypt using the software Jon wrote. In all probability the current version will be somewhat different but since we have the original code, reverse engineering shouldn't be too difficult. With that we need to start intercepting Globalstar calls to and from specific locations to find the source trying to kill us."

Jon picked it up again, "Michael, tell us about the phone number prefix."

"Remember the phone numbers called? As Jon pointed out stripping away the Globalstar code gave us a 916 prefix. There are three Mexican locations that use that prefix. One is near the Yucatan, another south of Mexico City, and Mazatlán. Guess who is near there?"

"How do we capture or intercept these calls?" Katie asked.

Jon explained that mobile communications, including GPS, use what's known as the L-Band which is a specific range of frequencies. GSM cell phones operate at 800 to 900 and 1800 to 1900 megahertz. Satellite phones use specific frequencies depending on the network provider. Globalstar uses 1616 to 1626.5 megahertz.

"We need to purchase a powerful digital receiver, tune to that frequency range and begin recording. We need to be located near enough to where the calls are being made to capture from the same satellite as it passes overhead."

"That's amazing. Now I understand your real estate request," when the others looked puzzled Katie continued, "Jon asked me to research available real estate close to Mazatlán for use as a 'home base' he called it."

"And what did you find?"

Katie's fertile mind had already jumped three or four steps ahead, "I told Jon the safest place in western Mexico is Cabo San Lucas at the tip of Baja – essentially the same latitude as Mazatlán."

Michael interrupted, "I agree."

She continued with a lilt in her voice, "When we're safe and relatively normal again, as we will be, I think it would be very cool to have a getaway on a beach at the Sea of Cortes. In the meantime I can just see Jon and Karl bent over their computers hacking satellite phone calls while Michael translates the Spanish they will be using. As for myself I intend to tan and swim with Marie and Corinna with beach service available."

Jon started to laugh. Slowly the others joined in but privately wondered if Katie had lost it.

"Just like the good ol' days," Karl chimed in.

Corinna asked, "You're serious?"

"Absolutely"

"I'm intrigued; show me please."

"Of course."

Katie went to her computer and moved the mouse to wake it up. Then went to favorites and made a selection at the bottom, "Here's my first choice. Tell me when to click forward through the images."

They gathered around and saw what would normally only be found in fancy travel advertising: palm trees lining a beautiful white sand beach, amazing contrast of blue from sky to sea and in a corner frame a tile roof stucco hotel looking structure.

Corinna said, "Next."

For the next fifteen minutes Katie went back and forth describing the property. She explained there is a highway called the Corridor running from the town of San Jose del Cabo near the airport about twenty-five kilometers to Cabo San Lucas at the tip. About a third of the way is a resort called Palmilla – golf courses hotels and all. On the sea side of the highway is the resort of One and Only Palmilla, accessible through a security gate manned 24/7. To the southwest of this resort is Palmilla Estates, a private, spread out enclave of eleven estates, entered through a second security gate. The one she showed was near the end and very private with the closest neighbor over a hundred yards away.

Marie said, "Unbelievable, I've never been to a place like that."

Corinna added, "Michael and I vacationed there for a second honeymoon. We love Cabo."

Karl said, "What are we waiting for?"

Jon and Michael just looked at each other.

A few minutes went by when the ever practical Marie asked, "Is this a rental? Who would be the renter? How much does it cost?"

Jon spread his arms towards Katie urging her to continue her roll.

"I can't see us renting anything as that would require a name or names and a vetting process. You may or may not know that Jon created a company during our previous activities that could be the buyer under some pretext, such as rental or investment property acquisitions. The price tag for this little beauty is $4,950,000 USD. I doubt it's negotiable. By the way you all know we deserve it."

Jon laughed again at Katie's mirth. So did Michael.

Katie concluded with, "I think we all need a drink."

Chapter 6

They spent most of Saturday morning assigning various tasks. Jon explained that when he had penetrated the financial network of The Consultants he had set up and used an off-shore company named 45172TCI Ltd. where TCI represented the Turks and Caicos Islands southeast of Miami. It is a British protectorate. That account should be used for any purchases, real estate or otherwise.

Because of their Spanish fluency Michael and Corinna would fly to Cabo, thoroughly check out the property Katie found and if suitable, make the purchase. Jon would wire the money as needed. They also would purchase a large SUV. The Lear charter would leave the following day.

Karl was given the responsibility to research and obtain the best available L-Band digital receiver with an antenna capable of capturing the low earth orbit satellite signals.

Katie would scour all available sources, including the Arizona Star files and archives and put together a package detailing every known aspect of the Sinaloa Province drug business.

Marie was put in charge of materials, supplies and logistics. Jon would obtain their own Globalstar satellite phone and get it activated, preparing to capture a call.

Now with a purpose and focus, they swung into action but didn't forget the need to be vigilant and take advantage of the security team supporting them.

The Lear charter, with Michael and Corinna on board, took off at 9 AM Sunday Tucson time. The flight time was estimated to be just over two hours. Katie had contacted the real estate selling agents Saturday afternoon, claiming to be the administrative assistant for an international investment company. One of the principals wanted to examine the property and if found suitable, discuss a potential purchase.

Any skepticism of the agents evaporated when Katie explained they were planning to arrive by company jet the following morning. The agents readily agreed to provide transportation service from the airport. Katie requested the agents make a reservation for two at the nearby One and Only Palmilla Resort – open for a number of nights.

Once at cruising altitude Corinna turned to her husband and asked, "What do you know about Jon's finances? He seems to have no qualms spending money. It's not something we're used to."

"I know both he and Karl are wealthy men because of stock in their public company ACT or Advanced Cyber Technology. Last summer when I needed funds for the stings and in particular the purchase from Zambada, the amounts needed were no problem and immediately available. I observed several transactions and they were Jon's personal accounts. Why are you asking now? If you're concerned, I have no problem raising the issue at some appropriate time."

"Thanks, just curious I guess. This is not something I could ever imagine."

The Lear landed and taxied to a private hangar area. They could see a white Suburban with Snell Real Estate on the door parked nearby. They deplaned and were greeted by the agents. It was half an hour later when they exited the Corridor Highway and turned left. They slowed and were waved through an arched entry by a security guard.

Karl quickly learned the very best satellite receiver for their purposes was manufactured by a German company named Radiometer Physics. Their L-Band receivers are somewhat larger than expected, standing over three feet high by the same in width, with the antennas, solar panels, seek and position sensors plus the control units. It was not too surprising. Karl found a US representative but with no units in North America. He was told one could be shipped via air freight, too large for either UPS or FedEx, on Monday. Karl told them he would call back.

Next he searched the Internet for information about the orbits of Globalstar satellites and their operating methods. What he found was mostly generalities although the method of 'handing off' calls between satellites as they move into and out of range was useful. He began to think about other sources, when an idea hit. He remembered a contact named Randy from ACT's Pentagon days that had moved over to NSA.

Karl's phone could connect to the ACT network over the Internet and he located Randy's email address in short order. He typed a short message and sent it, not expecting a response until tomorrow, Monday.

To his surprise his phone rang ten minutes later.

"Hi Randy, thanks for calling, remember me?"

"Of course Karl, what's up?"

"Before I impose tell me what's going on with you."

They chatted and played catch up for a few minutes when Randy said, "You obviously called for a reason."

"Realizing there are probably over a thousand satellites and other space junk up there, I surmise someone must keep track of it all. I need to know if the orbital data for low orbit communication satellites is available in some non-proprietary database. I thought you might know. This is a current project and we have chosen Globalstar as they have relatively few satellites. We did a job once for Hughes but I couldn't think of anyone there who might know."

"Yes the information is kept. We can pinpoint the location of any manmade space object at any instant in time. Do you still have your security clearance?"

"Yes."

"Send me a written request on ACT letterhead by email – for the files – and I'll see what I can do. Nice talking with you. Take care and I'll be in touch."

"Thanks." They both hung up.

"That sounded promising." Jon commented from across the room, "I've ordered the phone online. It will arrive by UPS on Tuesday morning."

Tuesday couldn't come soon enough. They were like expectant parents. On Monday Michael and Corinna spent several hours touring the property with the agents; then had a late lunch with them before initiating a possible purchase discussion. They explained the potential buyer's corporate position and registration having been briefed by Jon. The real estate agents countered with the sellers' position, a nonnegotiable price with no contingencies including mortgage financing. The agents went on to ask if Mr. Ortiz was familiar with Mexican Title Laws and the registration requirements for foreign ownership of Mexican real property.

Michael said he presumed nothing so would they explain exactly what is required. The female agent pulled out a dozen printed pages including forms and handed them to Corinna.

"You should look these over carefully. We have an attorney on retainer who can answer any questions. We cannot bring an offer to the seller from a non-registered entity or an unverified citizen of a foreign country.

"It is a pleasure to meet you; please call us," she concluded handing out her card.

Back in their room Michael called Jon and described the property as amazing and perfect. Then what had happened with the agents and requested Katie's fax number. From the lobby business center Michael faxed the pages.

Karl ordered the digital receiver and waited for Randy's reply. There was nothing left to do. Katie and Marie returned to the Star to organize and write up their research. Even Jon was somewhat bored until Michael's fax arrived.

Katie's email dinged on Tuesday morning while they were on their second cup of coffee. It was from Michael with a written description of the property. They gathered around the computer and read together:

'The Internet pictures didn't begin to show the magnificence of this place and in particular its functionality. It's over 8000 square feet, right on the beach with a large pool between the main house and those palm trees. The two story main house contains a large great room, dining area, kitchen, pantry with storage, powder room and an enormous laundry room. Why so large, I don't know. Upstairs are three rooms currently outfitted as an office, a guest bedroom and a library, plus a bathroom. Two wings on either side of the main building contain two apartment-like spaces. Each has a large bedroom, bathroom and sitting area. Simply amazing! The seller informs it's to be sold as is including all furnishings. How do we proceed?"

"Wow!" exclaimed Marie and Karl almost simultaneously.

Jon had completed and scanned all the forms Michael had faxed, scanned a variety of documents associated with the 45172TCI Ltd company and assembled them into a single file to be emailed as an attachment. "Are we good to go?" Jon asked the group.

He received yesses and affirmative nods from everyone. He then called Michael, who answered immediately.

"Great job, you two. Please make an appointment with that real estate lawyer – today if possible. Get me his email address and alert him to an incoming email with a large attachment of documents. They have our registered agent's signature scanned in and are set up for you to cosign. Let me know and thanks."

The buzzer at Katie's front door sounded. She opened the door to see a wide eyed UPS delivery person alongside one of the security personnel. She told him she would sign for Jonathan Reynolds, took the package and closed the door.

"Here you go – another toy for the boy."

Jon opened the package, extracted the phone, pressed ON and said, "Damn, more waiting. The battery needs charging."

Katie was starting to distribute and summarize her research when the email dinged again. It was Michael reporting an appointment at 3 PM and they were on Mountain Time. He added the lawyer's email address.

"Give me a minute sweetheart, while I send this."

Jon then used her computer to log into a seldom used email account he had set up earlier in the year. It would appear to come from the TCI Company. He attached the document file and typed a quick message of introduction and purpose. He signed it JR, and sent it. They were notified the forms were complete and correct.

Everything went smoothly except for a near battle between Jon and Katie plus Karl and Marie. On Wednesday Katie announced while Jon and Karl were gone to pick up Michael and Corinna, she and Marie would be having their hair cut, colored and restyled.

Jon burst out, "You're cutting your hair? No way!"

"What the hell did you just say?" Katie replied angrily.

"I can't believe you would cut that long gorgeous hair. It's as dark as your eyes. I can't imagine anything different."

Jon knew he had stepped into it but was being stubborn.

Karl piped up, "And what do you plan to do, Marie?"

"Alter my appearance," was her short retort, "and I don't want any comments from you like we just heard, understood?"

"Let me lay this out for your overworked technical brains. We know exactly what we're doing. We're not going to be holed up while in Cabo. Five minutes after we arrive there, someone will recognize me and probably Marie too. Five minutes after that the whole fricking world will know where we are. Did you think of that, boys? No further discussion allowed and if you don't like the result, tough shit, it does grow out. As for you two, no disguise would have much effect. I suggest you dress like tourists and wear hats plus sunglasses. Marie and I will also wear sunglasses whenever out – those large, very expensive Hollywood types. Have we reached an understanding?"

Marie could hardly suppress a grin. Jon and Karl just stood there dumbfounded. They were at least smart enough to keep their mouths shut.

On Thursday midday the Lear landed and Michael and Corinna were driven to their home to pack personal items. The purchase was complete. The registration documents had been filed. A full price offer was accepted to the elation of Snell Real Estate. Jon had wired the funds and closing took place two hours later. The villa or estate or whatever was theirs.

Upon Jon and Karl's return they found Katie and Marie, or was it them? They were packing cardboard boxes from piles of items Marie had purchased. If Jon and Karl had been dumbfounded the night before, it was nothing compared to their reaction now.

Katie's hair was essentially the same color but with many highlights. It was at her neckline and cut away from her face in a wild array of flips and curls. Jon was in shock.

Marie's hair had been somewhat straightened and cut into a bob style with long straight wisps. The strawberry blonde was now a real blonde.

Karl was in awe but recovered first, "What is your name beautiful, I'd like to meet you."

Jon walked up to Katie, scooped her up in his arms, held her tight and whispered, "God, how I love you."

They all burst out with laughter and giggles.

The plan was for three trips from Tucson to San Jose del Cabo over the weekend, one with each couple because of the amount of gear being transported and the desire for quiet entry. They knew from the pilots that arriving on a private jet and at the private hangars, customs would be cursory but nevertheless, caution should be exercised. Michael and Corinna would be first with their personal effects plus the large receiver. They had purchased a Ford Expedition currently located near the private hangars. Next would be Jon and Marie with computer equipment and half her supplies. Last would be Karl and Katie with all the rest of the supplies. They hoped to be settled by Monday at the latest.

There was a Home Depot with a garden center a short distance away. Marie and Corinna, with her command of the language, went there and purchased a wide variety of plants as large as would fit in the Expedition plus several shovels and watering cans. Upon their return they were planted around the digital satellite receiver. Everyone was satisfied with the disguise. Karl's contact Randy had come through with the ID, timing, azimuth and elevation of the Globalstar satellites. The receiver was optimally positioned facing south, southeast. It was powered up and connected via Ethernet cable to the computer in the office they rearranged in the main building on the second floor. The cable run had been disguised and looked like a TV coax.

Michael and Katie were given the task of grocery shopping. They were pleased to find a Sub-Zero refrigerator and separate large freezer included. The shopping list was long and included all the necessary makings for margaritas. Jon and Karl pitched in to unload when they returned but were banished when it came time to put the groceries away.

They discovered the restaurant in the nearby Palmilla resort catered, so with everyone tired, they ordered in a variety of traditional Mexican dishes. Karl descended the stairs from the second floor office to obtain a drink for himself and Jon.

He announced, "We're almost ready."

But just then the food arrived.

During dinner Katie addressed Jon who was devouring a shrimp dish with his head down, ignoring the others, and obviously in thought.

"Jon, give it a rest. Tomorrow is soon enough and we haven't finished unpacking and organizing all the supplies. OK, love?"

"OK."

Chapter 7

It was almost as if no one had a care in the world. Facing the Villa with the Sea of Cortez at their back, Michael and Corinna had chosen the apartment at the far left. They were sitting outside their sliding door on a small patio sipping coffee with sweet rolls. Next to them Jon and Katie had moved in and were currently unpacking. On the right, the inside apartment had been selected by Karl and Marie. The far right, unoccupied, was for guests – whoever that might be – and for Stephanie over Christmas break should they still be here.

Marie had volunteered to prepare breakfast. She was currently getting used to the kitchen and had decided French Toast would be simple to prepare. After breakfast it was time.

The control software associated with the receiver had been installed on a workstation Karl had set up in the upstairs office. He sat in front of its monitor and using Randy's provided information began to adjust the azimuth and elevation of the antenna, having already adjusted the frequency filter to the 1616 to 1626.5 megahertz range of the Globalstar network. Within a few minutes the ID's of several satellites appeared on the screen. Karl made slight adjustments until the signal strength was optimal. They had identified and captured the passing communication satellites. The next step was using the receiver's primary function: signal capture.

That functionality was invoked. A window opened and an oscilloscope type of signal appeared on the screen. It was intermittent. The entire assembled group watched in fascination. Further they watched in amazement as the main screen showed a change in satellite ID and the capture signal died from the original one and revived for the new one.

Karl commented, "It's pretty obvious we have a lot to learn and a ways to go."

Katie commented, "As usual I'm in awe of you two."

Marie added, "I second that motion."

They all retired except Jon and Karl who stayed at it until they became bleary eyed sometime after 3 AM. Despite that effort all were up consuming delicious sweet rolls with their coffee the next morning. Jon was still up in the office when Karl informed, "We captured several signal streams from the Globalstar satellites and Jon is working through the decryption process. Stay tuned."

By mid-morning they were growing restless and wondering what was next when Jon emerged at the top of the stairs and said, "We've got it."

He descended and explained, "The algorithm currently in use to decrypt communication signals is a variation of what ACT provided years ago. The architecture is essentially the same. The criterion is quite different."

"Explain in plain English lover," Katie asked.

Jon poured a cup of coffee, sat down and explained, "When a satellite phone user, or any phone call for that matter, initiates a call a transceiver responds to the electrical signal, then a switch passes the signal to a server that determines the source and destination, next authenticates it and sends it back to the transceiver to pass along, in essence, make the connection. That's a simplified version.

"What Karl and I just did is to call Katie's landline on the Globalstar phone, wait through the voice message response, all the while recording the satellite activity – understand it was part of other transmissions – and then recognize the breaks and separate them into parts. Using a combination of the transmission header packets we were able to identify and piece together our particular call. Then it was a matter of trying our decryption algorithm to decode the message left on Katie's answering machine. We haven't been able to do that yet, but hopefully real soon."

All but Jon and Karl went to bed. At about 2 AM they had done it. Bit by bit they reverse-engineered the encrypted, captured call based upon the original source code Jon and ACT had written together with experience, intuition, and a lot of trial and error. They both looked at the transcribed text on the screen. Despite several punctuation errors, it was verbatim the message left on Katie's answering machine. A high five was exchanged.

Karl emerged the next morning, sufficiently sleepy, looking at the others, and said nothing while he fumbled for a cup of coffee. Shortly thereafter Jon went through something similar.

"Wake up boys and tell us what's happening. Daylight's burning," Katie said with no mercy.

Karl responded, "We're able to decrypt the calls." Jon simply nodded.

The enormity of it all began to sink in.

Jon addressed the group, "Here's the situation. I need to write a script and some lower level code today. That will automate the process. The receiver will operate 24/7 receiving a data stream from one satellite after another as they pass overhead. It will be a nearly continuous stream. We only have to recognize and decode the header portion as that contains the receiver and sender phone numbers that the network manages. The data stream is not continuous, nor are the individual packets associated with a particular call all together. It's somewhat more complex but for our purposes now, should a header packet contain a call either sent or received with a 916 prefix, we will capture all associated packets while discarding the others.

"Remember why we're here: the phone requires line of sight to connect with a satellite and the same ones we will be monitoring will almost certainly be the ones receiving and processing calls from anywhere in Sinaloa Province.

"When the last packet in the stream has been captured and appended to the preceding ones, we will run the decryption process and see what the call was about. I'm sorry to say there will probably be a lot of chaff with the wheat."

"What would you suggest each of us do while this process, as you call it, is running?" Katie asked.

Jon was tempted to remind her of the beach and swimming comment but thought better. Instead he offered, "Monitoring will be an around the clock effort because we don't have unlimited disk space and this capture produces very large files. I suggest you, Marie, Karl and I take six hour shifts. Michael and Corinna are needed to read the call text, translate and identify calls of interest. Remember also there are many 916 prefix phone numbers. How many use satellite phones, we don't know."

"When does this get underway?"

"As soon as my script is finished, that's hopefully in a few hours."

"Since I'm a night owl," Marie offered, "I'll take midnight to six."

They quickly established a schedule and Jon went up to the office to do his work.

Guzman felt safe enough to leave his eastern Sinaloa underground hideout and return to his Mazatlán safe house where he met Zambada. Creature comforts have their draw. Slowly his business was returning to normal – the new normal.

His focus had returned to the chaotic events of the past almost five months. He asked without addressing anyone in particular, "Who has been chosen and dispatched to eliminate the rest of them?"

Everyone knew who he meant.

His primary lieutenant answered, "Ernesto Guevara has been pulled from security detail and is in Seattle. I expect to hear from him later today or tomorrow."

"I want it done and over – maximum publicity – understand?"

"Si, Jefe."

They were up and running by 6 PM. Katie had volunteered for that shift. It took a while to work out the procedure but soon it was running fairly smooth. Whenever Jon's script identified a call, incoming or outgoing, where one of the numbers had a 916 prefix, a message box appeared indicating recording was in process. Upon completion a file name with a date-time stamp was displayed in another message box. It was the responsibility of the shift person to run Jon's decryption program on this file and log the new file name containing the call text for either Michael or Corinna to read. It had been decided they would take six hour shifts: 6 AM to 12 Noon and then to 6 PM. Jon or Karl during their respective shifts would review the translated analysis and comments.

Awake and off shift persons would prepare food, usually snacks for those on shift. After two days, there were a surprisingly few number of calls with that prefix. Jon suggested most legitimate people with that prefix would be using land or cell phones, not satellite. Then there was a hit.

Corinna read the text but could not make out the words. She awakened her snoozing husband for a look. He couldn't either. It was a message left in a language he did not know.

Later after Jon had reviewed the hit, he asked, "What do you make of this?"

"I don't know. Let's listen and see if our voice-to-text translator is whacked out or something."

They did so and the voice was clear, but definitely not in Spanish.

Michael asked, "Do you recognize the language?"

Jon said, "No, but I know how to find out. There's a free app on the Internet. I've used it before."

Ten minutes later the language was identified as Pashto.

Jon said out loud, "What the hell did we just run into?"

Michael responded, "I have no fricking idea."

The phone message source was not recognizable, but they noted the destination phone number with its 916 prefix. Jon modified his script to include this number in all future recordings, in addition to the generic 916 prefix. The monitoring continued.

The next afternoon during Karl's shift, from noon to six, another hit occurred with the previous day's specific phone number. Michael translated almost immediately then listened to the voice decryption again. He rousted everyone.

"Read my translation carefully."

'Hello, Jorge?'
'Yes'
'Ernesto, here, no success here in Seattle, going to Tucson in the morning. That cop is so well protected I can't get close. Oberling's are nowhere to be found. House is dark and unoccupied. Locals are useless. Sorry no success.'
'I will report.'
'Alright … good bye'

All were stunned and the many weeks of fear returned.

Jon asked, "Are you saying that call was made to the same number as yesterday's Pashto message?"

"Yes"

"I'll add this caller's number to the script. Sorry but we need to keep going."

It was Sunday and they had been at it for almost five days. They decided to let the software run on automatic while they had breakfast in the sun by the pool. The strains were showing.

Katie decided to break the silence, "We have had seventeen calls to and from that number. We have identified eight other numbers. We all know instinctively this is another spider's web just like before. We are pinpointing communications between key players and their leader within the drug trade. I for one am overwhelmed and very concerned. Please don't tell me we're losing sight of our purpose – personal safety, privacy and a normal life."

The ensuing silence could be cut like butter.

Chapter 8

Jon and Michael were huddled by the pool. The others were losing their tensions on the beach and in the clear saltwater. Overall it was a remarkably resilient group of people with Katie leading the way.

Jon began, "Michael, what I'm going to tell you is out of character for me. I'm angry to such a degree it's starting to control my thoughts and actions and we both know that's not good. Nevertheless I want to send a direct message followed up by another. First trust me that I know how to pinpoint the exact location of that Ernesto asshole at a point in time. If I could orchestrate it, I would have my own shooter there to put a bullet through his brain. But unfortunately while I can locate him, I can't do it ahead of time."

"I do know what you're talking about – GPS. I can relate to your feelings. As a cop I've been trained to restrain, as they say. From the call we know he cases possible locations. Surely he will case Katie's condo, my house and the Star office. Suppose we place either cops or security at those locations and make the call. If he resists hopefully they will shoot him."

"Set it up Michael. I can't thank you enough for your friendship and support."

"Just make sure you take the very best care of my surrogate daughter Katie, and you're welcome."

Michael spent several hours making over half a dozen phone calls. He told Jon it was ready to go. They would either have to guess the timing or wait some hours for another phone call. Their timing calculations put Ernesto in Tucson about six hours before Michael's ready to go report.

Katie yelled down from the office. It was her shift, "We've got another call now."

They all rushed upstairs.

It took over a minute for the software process to run and Michael, with Corinna over his shoulder, to make sure of the translation. It read:

'Jorge, it's me Ernesto'
'What is your situation?'
'I'm in Tucson and just like in Seattle there is no one at the Ruiz woman's house. It is dark but there is a car in the garage. I'm going to that cop's house next. I will check out that newspaper office in the morning.'
'Report then. I will not pass this along for now.'
'Bye'

Michael asked Katie, "Best guess from your place to mine?"

"Twenty-five minutes this time of night."

Michael got on the phone immediately and told Jon to get ready to make his call. He hung up and further instructed Jon to make the call in twelve minutes. Karl brought up the software application that would determine the caller's and receiver's GPS locations. He reported ready to go.

Michael called back and was patched to the lead squad car.

"Cody, is that you? I'm happy you're there. Keep this line open." Turning to Jon he said, "Make the call."

Jon punched the numbers into his satellite phone and waited. Within thirty seconds the software captured the header packet of the call. Another thirty seconds later the call was answered.

"Quien es este Jorge?

Jon responded in English, "Sorry, wrong number," and hung up.

Karl provided the GPS coordinates and Michael had Cody enter them into his squad car console.

Cody came back with, "He's about a quarter mile from us; we will deploy. I will call back."

Michael said, "Good hunting."

They waited almost half an hour until Michael's phone rang. He put it on speaker.

"Sarge, the perp is dead. He pulled up into your driveway and got out. An officer you know, Ed Ferris, stationed next to your garage ordered the perp on the ground. The perp tried to pull a gun – we found the gun later – and that was it. I think his weight went up five pounds from all the lead; any instructions Sarge?"

"Cody my young friend, I think you deserve to be Captain someday."

"Thanks Sarge, be in touch."

Jon picked up the SAT phone and made another call. After five rings it was answered.

"Who is calling this number?"

"Your worst fucking nightmare and I know you speak English. This is Jonathan Reynolds calling. I also know you recognize my name. Listen carefully asshole. Your hit boy Ernesto is dead and full of lead. Hey asshole, that rhymes. Send another and we will kill him too and then bring our attack to your door, comprende?"

Jon hung up.

Katie began to clap her hands. The others joined in. They expunged the built up tensions.

Jon's adrenaline was still pumping and he didn't join in the celebration. When things became normal again he said, "We scored a run but I don't think the game is over by any stretch of the imagination. They won't go down easily."

"However we know where they are right now. That's really something," Karl responded pointing at the computer screen.

The GPS tracker application had been left running. It showed the coordinates of Jon's last call.

"Incredible," said Michael, "what can we do with that information?"

Guzman's face was flushed with anger as he spewed one string of expletives after another. His lieutenants stood nearby in total shock.

"They got Ernesto's phone. Get that shut down right this god damn minute. And I want all new numbers."

One of the lieutenants responded it would be done by morning.

"What are we going to do, Jefe?"

"I'm thinking. This is not over. Nobody fucks with El Chapo."

Jon brought up Google Earth, clicked the GPS option and entered the coordinates of the call. The screen rotated and zoomed in over Mazatlán with a red pin in the main city. Jon zoomed in further as the others watched. He kept going until the resolution became fuzzy. On the sidebar he clicked the *Roads* option and zoomed out slightly to improve the resolution. They could clearly see a yellow house with a red tiled roof along with the nearest major roads.

Michael repeated, "Incredible, I wish I could zoom there myself, right now. I'm tempted to call that Mexican Army guy we worked with during the incursion last summer."

"I'm sorry to say that involving authorities or governments at this point would not end this and probably make things worse," Jon responded.

"Unfortunately, I agree," added Katie.

"We just issued our own threat, much like the email. We need some serious leverage to bring this to an end. I think we should keep up our monitoring a while longer and see what shakes out."

Michael went on, "I'm amazed at what you and Karl have done but I don't understand why the US and Mexican governments or departments within aren't doing the same thing."

Katie said, "Let me answer. I included that in my research package. Short answer is they do, all the time, and in particular the DOD, DEA and NSA. Their primary focus is land based and cell phone based communications. It occurs in both the US and many other countries. I asked Jon the same question. Land based telephone communications are not encrypted so monitoring is based on making a deal with the providers such as AT&T; GSM or cell phones are encrypted but use a very weak algorithm. Right Jon?"

He nodded.

"The seeming ease of monitoring becomes complex or overwhelming because of the sheer volume. Realize there are many billions of calls every day, so how could they identify terrorists or drug deals or whatever criminal slash national security areas of interest. They can't without some detailed, prior knowledge to focus upon."

Jon picked it up from there, "Now we come to our situation – satellite phones. There are a fewer number of providers; each uses a proprietary, strong encryption method not in any government's control although I'm sure they're constantly trying to hack them. Lastly the phone providers don't vet anyone – buy it over the Internet and activate. It's as simple as that. I used a phony name. The numbers are only known to the phone user and the provider and there is no fixed location such as your home address or a specific cell tower. To monitor satellite phone usage you must do what we're doing and without that captured phone we would be nowhere. Actually we were lucky to see the recent calls. Note that the phone was disconnected almost immediately."

They resumed their schedule.

They were collecting seven hits a day on average with two different numbers, both with a 916 prefix. Outside his shift Jon created a database to organize the hits. It contained a date and time, caller and sender numbers, a blank column for the location of non 916 numbers and a synopsis of the conversations. He couldn't capture any GPS coordinates unless he made the call. A pattern was developing and clearly most of the calls' subject was directions given and reports received during the conduct of a drug trafficking operation.

He had caught up with the calls log they kept. There were now sixty-eight calls in the database. He scanned the data and noticed two outgoing calls to the same number that left the Pashto language message. His mind wondered about current technology for voice translators – not in his area of expertise. He began various Internet searches on the subject and discovered there were quite a few applications available. The reviews on each were very favorable as to accuracy.

He returned to the database and read the synopsis for each of the calls. It had been written by Michael after his Spanish translation. The first requested clarification of amounts and timing. The second stated payment had been made.

Jon briefed the group about his database and his discovery. He asked Michael to review his translation in light of the certain Asian connection. He asked Karl to research voice translators both real time products and post process products and additionally to obtain the best available.

The next day during Katie's shift she hollered for Michael and Corinna to come to her computer. She brought the Spanish text for them to read. They immediately translated.

'Greetings El Chapo'
'And to you El Mayo, what brings you?'
'I may have a solution to the ephedrine problem.'
'I'm expanding my Panama transfer point from Peru and Bolivia. The Colombians you know. It will include a warehouse at the wharf. Consider it.'
'I will do so. Do you still intend to be here in January?'
'Yes and I'm very pleased with the shared routes.'
'Good, later.'
'Alright, later'

"I'll be damned," exclaimed Michael.

"Wasn't that who you dealt with on our sting?"

"Yes."

Corinna asked, "What's this ephedrine problem?"

Katie answered, "Ephedrine is one of the main ingredients used to produce Meth. It usually comes from China or India. The Mexican government recently banned its import. This just gets more and more intertwined."

Jon entered this call in his database and announced, "We have identified a new Guzman phone number. The other end of this call matches another one we have and now can put a new name on it. I have half a mind to call the asshole again, but will restrain."

During Karl's shift he purchased, downloaded and installed a voice translation application using a credit card Jon provided.

"Hey, Jon and Michael com'ere would you? I have that app you wanted installed. Let's try it out on a few of the Spanish conversations."

They ran the voice track of two and Michael was amazed again, "Guess I can go home. Is there anything you two can't do?"

"C'mon friend it isn't perfect so no, you can't go. Let's try this on that Pashto call."

They did and were shocked at the message left.

'Sin666, Pak11 here
'90 MT ready, 4kusd each
'Pay need
'Leave K next day
'Reply'

Karl said quietly, "We're caught in some kind of vortex. It just doesn't end. Is there such thing as a simple, private life?"

"Speaking as a cop, I don't think so. What are we looking for Jon? Are we going to find it? Or as Karl says get sucked into the vortex."

Chapter 9

As Jon produced the code, Karl tested it. Within five hours they were finished. The entire satellite monitoring process had been automated, from signal capture to translated text, with everything in between copied and stored. Even the database was being populated automatically.

The others were outside on the beach. Jon and Karl went to the pool and sat down to watch. It almost seemed they were vacationers without a care in the world. They all knew differently.

Soon a catered lunch was delivered from the nearby resort and they gathered around the pool to consume it. Jon told them what he and Karl had done so everyone was off the hook. Only a daily check and review would be necessary. Maybe some fun and normality could be had – at least here.

Karl and Marie called Stephanie every day. She was lonely but well. They asked the group if anyone had given a thought about Christmas. It was less than ten days away. No one responded but the various looks said it all: surprise, sadness, wonder.

Ever the mood manager, Katie suggested, "As beautiful as this place is, maybe a Christmas break could happen."

Marie said, "Actually I was just wondering about your thoughts. Karl and I have considered flying Stephanie and my sister's family here for Christmas, but we wouldn't want to interfere or impose on any of you."

"What a great idea Marie," Katie exclaimed.

"I totally agree," added Corinna, "but I would like to be home for at least a short while."

Michael added, "We've been invited to Ruben Valenzuela's estate for Christmas. We accepted before all this began and haven't followed up. Is that what you're thinking my dear?"

"Yes"

"Where do you want to spend Christmas, Jon?"

"I really don't know – haven't thought about it – your choice."

"You're not much help."

"Sorry, right now I want to call Guzman's phone direct and also that Pakistan number. I can't seem to rest. After everything we went through, I'm thoroughly disgusted and major league pissed. Underneath, sometimes anger management is not my strong suit."

Katie came over and sat in his lap, hugged him and kissed his cheek. He held her tightly.

"Follow your instincts, lover."

Guzman was eating an early dinner or late lunch, shared with his most trusted lieutenant. Periodically he thought about how to take out the reporter and her friends. Usually it was too simple. The two failures and the call from Reynolds had shaken him badly and that was unusual. He was lifting his fork when his phone on the table rang. There was no Caller ID.

"I'll get it," answering, "hola."

"Jonathan Reynolds calling to make certain you received my previous message and will be heeding it."

"How, where?" sputtered Guzman.

"I'm watching you. Do you feel safe in Mazatlán in your yellow house with the red tile roof? You shouldn't. For all you know I'm just verifying your location so I can send a drone to explode and kill your sorry ass. Don't think I can? Think again. Here's the deal, asshole: I could provide your location to the Federales, or the other Cartel butchers, or simply kill you myself, but I won't if you forget about us. It works both ways. I won't wait for an answer because you're smart and want to survive, Ciao."

Guzman's hand was shaking and he dropped the phone on the floor. The lieutenant was dumbfounded.

"How much of that conversation did you overhear?"

"Most, you had the volume up."

"Forget it ever happened. I want all new phone accounts and phones, not just the numbers changed. Destroy every one we're using now. And get our technical people on how this was done. Do it!"

"Si, Jefe."

As Jon was hanging up he turned to the group and said, "I can rationalize making that call instead of alerting DEA, Mexican Police or whomever because by the time they got there he'd be gone, just like before."

Michael said, "I really believe you did it."

"Me too," added Marie, "I can feel myself starting to relax."

Katie said, "You're one scary dude, Jon Reynolds, but I love you anyway." That generated a chuckle all around.

"What about Pakistan?" Karl asked.

"Tomorrow, I want to sleep on it."

And sleep he did but not until after making love for the first time since coming to Cabo.

Katie was swimming laps in the pool when Jon emerged after ten o'clock. She waived and kept up her pace. The others except Karl were at the beach. Jon sat down next to his longtime friend and said, "Morning."

"That it is; that it is."

"Did you check our process logs today?"

"Yes and there's another exchange in a new language, it's in an Afghan dialect of Dari. That voice translation app is amazing."

They went inside and up to the office for a look-see. They read the translated text, Karl for the second time.

"It seems someone over there is very pissed off because the payment was short. Reading between the lines, Guzman contracts with a Pakistani distributor and shipper who obtains heroin from these people in Afghanistan. Somebody stiffed somebody and Guzman claims it wasn't him even stating the payment date and amount. The next exchange is really explosive when the Afghans realize the price difference. Guzman confirms. I guess markup is not understood in that part of the world," Karl chuckled.

"What a screwed up world we live in."

"Well at least it's on the other side of the planet."

"Karl, I can't believe you said that. How many of our troops are there? What about 9/11?"

"Sorry, it just caught me as funny how crooks and normal businessmen all try to screw each other."

As everyone was seemingly relaxed with the perfect weather and surroundings, Jon and Karl chose not to report what they had seen. All the girls were going into Cabo San Lucas, to shop, check out the Cabo Wabo Cantina and would return, 'who knows when'.

For the first time Jon took a swim in the pool, then the ocean and back to the pool. Even he was starting to relax.

Two days later, and less than a week before Christmas, plans were laid. While enjoying another delicious catered dinner, they finalized the plans. Jon would send the chartered Lear to Seattle and bring back Stephanie and Marie's sister plus family. Once they were settled Jon and Katie plus Michael and Corinna would fly to Tucson. Michael had confirmed with Ruben Valenzuela and he had extended an invitation to Jon and Katie; he was most insistent. The return flights would occur two days after New Year's 2007.

"Jon?" Michael asked after dinner, "All these jet charters, this incredible place, various expenses and so on. We have put up nothing. I understand our various deals last summer and your financial support, but Corinna and I are more than curious, maybe almost uncomfortable, with how this is being funded. Care to share?"

"You knew this was coming, big boy," Katie chimed in.

"Yes, you deserve to know. Sorry I've been remiss as it is the least important thing on my mind. No secrets among us. All funding, from the beginning, has come from my personal accounts except for a few credit card purchases I didn't want connected to me. That includes this place and the jet charters. Both Karl and I are fortunate to have been on the ground floor of evolving, in high demand, technology that propelled Advanced Cyber Technology to explosive growth and a public offering. I don't mind telling you the current value of my ACT stock is over fifty million dollars.

"Some of you may remember the description of drug seizure laws outlined by the Pima County DA, Wes Davis. Following those criteria and during the hack and destruction of The Consultants' New York facility and computer network, I decided we needed to collect our own consulting fee. Those funds currently reside in several accounts outside the US under the name of the company we're using here. They total just over a hundred million dollars.

"When the dust settles on all the prosecutions and seizures, which will probably take a year or more, I expect the seizure laws will justify those funds. Should that happen each of you are equal members slash partners in 45172TCI Ltd. In the meantime I will continue to fund. Rest assured 45172TCI Ltd will get my invoice for what I've spent – sufficient explanation?"

The look of shock was apparent on every face except Katie. She had not been privy to the numbers but was not the least bit surprised.

Part 3 – A New Perspective

Chapter 10

The blindfolded children, twelve in all, were wielding their sticks at the piñatas suspended from low spreading tree branches. The largest one, star shaped, was in the center of a large neatly trimmed grass area. The smaller ones depicting various animals such as donkeys –some were difficult to discern the species – were spread out on outlying trees. Otherwise the sticks could have inflicted an unintended hit.

The adults seated around the grass area gave clues as to how close one was and applauded whenever a hit was made. The children squealed in delight when their hit dislodged presents that fell to the ground.

It was Christmas Day 2006 at the hacienda of Ruben Valenzuela, south of Nogales, Sonora State, Mexico. In addition to the families of Valenzuela's sons and daughters, the invited guests were Michael (Miguel) Ortiz and Corinna, plus Jonathan Reynolds and Katie Ruiz.

Valenzuela had been instrumental in providing key contacts and information, specifically Zambada, El Mayo, for Michael that allowed Jon and Katie to unravel a worldwide financial network of illicit drug trafficking proceeds and assets.

The blindfolds came off; the children scooped up the many presents on the ground and happily ripped off the wrappings to reveal the treasures within. The smiling and applauding adults, to varying wistful degrees, remembered their Christmas childhood memories. The feast was soon brought out and placed on several long folding tables. The chili rellenos and churros were the hit of the tables.

Jon and Katie were sated and nursing an after meal Margarita when Ruben Valenzuela approached them.

"My family is pleased and proud you accepted our invitation. It is a great pleasure to meet you. Miguel has told us so much about you and we actually do have television in this remote area that followed your path. Welcome and salute. Perhaps we could have a brief talk when los niños are to bed."

Jon stood and shook Valenzuela's hand and said, "Our thanks to you for everything. You have a beautiful family."

Katie noticed Ortiz in deep conversation with one of the Valenzuela sons while his wife was left alone. She rose and went over to correct that situation and visit. Jon's eyes passed from person and group taking in the entire throng. His attention then focused on two young people standing back at a distance without the happy, holiday cheer expressions of the others. They were male and female but amazingly similar in looks. They might have been 17 to 19 years old. Jon had caught their stares several times that day and it made him curious.

They were gathered in the great room of the hacienda. Candles burned from eight wall sconces. The leather chairs and sofas formed a near circle around a low, tiled water feature with a bubbling inlet and many small koi could be seen swimming past the underwater lights. Two of Valenzuela's sons served warmed brandy from silver trays. The other sons and daughters were putting the grandchildren to bed. Michael and Corinna Ortiz sat side by side on one sofa, Jon and Katie on the other. The two sons and Ruben took the single chairs.

Michael leaned forward to propose a toast, "To our hosts – Feliz Navidad and thank you for your hospitality. Let us toast the most courageous and honorable man I have ever known: to you Senor Valenzuela."

"To you," Jon and Katie said in unison.

"De Nada, you are most welcome," Valenzuela replied. He went on.

"I am happy our family could be of help to you. It is remarkable what you accomplished and I wish I could be more optimistic your efforts will affect a long term solution, but alas many of our young people seek only instant gratification.

"In addition to enjoying your company, the reason I asked for a talk is to convey my concerns and suggest your constant vigilance. With governments everywhere seizing the illegal drug money, the drug traffickers themselves running for cover or being captured and prosecuted, the supply routes disrupted, and everyone in a feel-good mode, this should be a time for reflection and caution.

"Senorita Ruiz … Katie, said it best: with no plan to deal with the drug demand we will repeat the past. I do not see such a plan forthcoming. Furthermore the remaining Cartel leaders are angry and will seek revenge. That is their nature. At best we have a year or two before they heal their current wounds, re-align and re-emerge as deadly as ever. I for one plan on increasing the security on our estate and alert everyone, family, friends and workers to be vigilant. Please do the same."

Katie spoke first, "Senor Valenzuela, this year I have learned to trust my instincts and that includes warnings from knowledgeable people I trust, such as you. Rest assured we are forewarned and forearmed. We understand and respect any steps or precautions you take here on this magnificent property. Lastly, anyone who might attempt to bring harm to you or your family is our enemy."

Jon raised his snifter, "muy de acuerdo."

Michael stood up, "muy de acuerdo."

"Muchas gracias amigos," Valenzuela responded.

The intensity in the room subsided and the conversation turned to their mutual enjoyment of the children and the Christmas piñata activity.

Jon spoke up, "I noticed a pair of young people, very similar in appearance. They stood back from the group. I also caught their eyes on several occasions. Who are they?"

"Welcome members of our family. After the tragedy of 9/11 and the United States attack the next month, their parents, who had been professors at the University of Kabul, fled to

Pakistan. They tried to seek asylum in your country but could not. They also failed with Canada. They were multi-lingual and applied to Mexico. They arrived here a few months later. They were a very dedicated and brilliant couple and obtained appointments at the Instituto Technologico de Sonora. They died in a tragic accident two years ago.

"My youngest son Ramon had been one of their students in his last year and after the accident asked if we could help. They are twins and we have provided a place near Ramon ever since. They are good students, currently studying at Instituto Technologico de Sonora and as bright as their parents but do shy away from our family activities, I'm afraid. They still talk about living in the States."

"Thank you, just curious. I guessed them to be about eighteen."

"Twenty, next month I believe."

Ramon added, "More than bright, more brilliant than their parents. They were curious about you two because they asked me."

Michael and Corinna decided to stay another day. Jon and Katie stated they must return. They all said good night and vowed to maintain the budding friendships.

Re-crossing the border at Nogales was uneventful, but both Jon and Katie were ever vigilant. Valenzuela's warning was nothing new.

As Jon pushed slightly over the speed limit and Katie's head was resting back, she commented and asked, "What a fabulous day. It was the best Christmas I can remember and all the better being with you, lover. I hope you enjoyed yourself too. By the way I'm curious. What was your interest in those twins?"

"I'm curious too. I can't put a finger on it. I caught them staring at us. Both made eye contact on several occasions. We didn't acknowledge each other at all but there seemed to be some kind of connection or recognition. I can't explain it."

As an extra precaution Jon hired the previous security team again while they were in Tucson. Katie disagreed but deferred to Jon. She wanted her freedom of movement back without worry. Changes were being made.

Katie met with her editor Edgar Greeley, related to Horace as he claimed, and announced she sought a sabbatical from her weekly column syndicated through the Arizona Star. Before Edgar could start to convince her otherwise, she asked, "Please, I need it?" Edgar told her he would go to the wall with the publishers to support her. Katie left the meeting with great optimism.

Michael hand wrote his letter of resignation and intent to retire. He had been on the Tucson police force for twenty-five years. He personally handed the letter to the Captain. Everyone currently in the squad room watched as they knew about the letter beforehand.

"Michael, I'm reluctant to accept this but must respect your wishes and desires. You will be sorely missed around here. I don't know your future plans but … please listen … we have an on call, as needed, consultants program that I encourage you to check out."

The Captain stood and shook Michael's hand vigorously.

Michael said, "Many thanks; I'll think about it."

He left to clear out his desk and visit Human Resources to complete his retirement. Doris, the office administrative assistant, snoop, gossip and general nemesis to Ortiz over the years mouthed 'good riddance' Michael glanced her way and presented his middle finger. A number of officers clapped.

During the week between Christmas and New Year's Karl and Jon kept in touch with Karl reporting many interesting new entries in the database. He reported they could wait until Jon returned. He also reported his family was having the time of their lives; he hadn't seen Marie and Stephanie this happy in a long time. His sister-in-law's family was almost star-struck.

Jon and Katie took it slow and easy with several hikes and dinners at El Matador. The security details he had hired remained in place, including the Ortiz household and Edgar's travels between the Star and his home.

On New Year's Eve they joined a catered lunch at the Star, said goodbye to the newsroom staff and collected Katie's personal items. Edgar had damp eyes as he gave Katie a strong hug, "Now don't you become a stranger, young lady."

"I promise to keep in touch and many, many thanks for your help with the syndication and everything else you've done for me over the years. I will be ever grateful."

"You have deserved and earned it all. Bye for now."

Chapter 11

It was Wednesday, January 3, 2007. The estate at Palmilla was back to business as usual. Katie was typing away on her last, for a while, article; Corinna and Marie had volunteered to do bedding laundry and general cleanup; Jon, Karl and Michael were examining the satellite calls captured over the past ten days.

"What do you two make of all this?" Karl asked.

Jon responded first, "My business sense nose tells me there is much infighting between the heroin producers in Afghanistan and others in the supply chain. It seems Guzman is in the best position of all as he controls the receipt and distribution of the product and is the paymaster. At a minimum I would say there are three Afghan groups and two Pakistani distributors involved with Guzman."

Michael added, "I agree but my cop's nose found something else more interesting: the coded references to their facilities and routes. The communicators are very circumspect and well disciplined. It would appear that Routes One through Five are into the US. K refers to Karachi, P refers to Panama and S refers to either Sinaloa or Sonora or both."

"You guys must be asleep. You didn't even notice the phone numbers. They are all new, although the prefix is the same and the timing is very interesting. None of the original numbers we identified have been used since the day after Jon's second call to Guzman.

"And nothing between Chapo and Mayo; weren't they going to meet in January? As I enjoyed the holidays here and checked in on our process, the obvious questions popped up. What are we going to do with all this data and how can we be sure we're off Guzman's radar?"

"Good questions Karl. I'm not sure either, but I'm itching to locate those Afghan phones. You know there's a full scale war going on there right now. Would you call or email your NSA contact, Randy? I would like to know if they or some other department is doing what we are. You might allude to the fact we inadvertently ran across some Afghan communication in the course of your project."

"Will do."

Karl emailed Randy and took a return call on his cell less than an hour later.

"Happy New Year Karl, what's up?"

"Same to you – in the course of my project, and thanks again for your help – we ran across some very interesting communications that seemed to originate somewhere in the Middle East. I became curious if you guys might be doing monitoring of those low orbit communication satellites?"

"Before I answer can you share something about your project? This is a sensitive area not to be discussed on a cell call."

"We've been retained to update and test a new, much stronger encryption algorithm. Do you remember my mentioning our previous contract with Hughes?"

"Yes I do. What I can tell you: we have a small group focused in that area but the traffic is so low when compared with other networks and we don't control the algorithms being used, not much intel is gathered. In fact our sporadic attempts to reverse engineer the encryption have not met with any success. It seems the hash keys are constantly changing and random."

Karl chuckled, "You can thank my buddy Jon Reynolds for that. Remember him?"

"Of course, is he still with you? I followed all your activities. Reynolds and that gorgeous reporter are quite a pair. I'm not hearing the real reason for this call. Fess up Karl."

"This is a preliminary communication. As I said we have inadvertently run across some communications that might interest you. We have a bit more work to do. Would you be interested in possibly meeting and exchanging some information?"

"Now you have my attention. I probably should report this call took place."

"I don't have a problem with that, but any report must reach the desk of Preston Gates, the AG. He's the only one we trust. I will be in touch, probably sooner rather than later."

All together again with everyone having the same current information, a revised plan was made. Karl and Marie would return to Seattle as he was due back at ACT by the tenth. They would install a very sophisticated home and grounds security system together with random but frequent security patrols. Michael and Corinna would return home and get their house ready to sell. With their time in Cabo and now understanding their new financial future, the retirement they had always dreamed about was possible. They would also use the Tucson security team. Jon and Katie would stay at Palmilla, at least for some time. The satellite communication capture process would continue to run with them monitoring it.

Four days later they said their goodbyes and departed to their respective locations. The Lear would fly to Tucson first and then on to Seattle. Jon informed the Lear charter company that afterward, they were not needed until further notice.

Katie and Jon were splashing in the small surf like a couple of kids when they heard one of their cell phones ring on the table by the pool. They ignored it.

After drying off and fetching some lemonade with a shot of Tequila each, Katie picked up her phone. The missed call showed a number with a 202 area code she did not recognize. No message had been left.

"Think I should call back?"

"Up to you but if someone is trying to reach us you'd think they would've left a message."

Ever the curious one, Katie pressed Call Back.

"Preston Gates here." Katie's phone did not broadcast Caller ID, simply Wireless Caller.

"Mr. Gates, it's Kate Ruiz. I saw a call from your area code and returned it. Happy New Year, how are you?"

"Very well my dear and you?"

"Good."

"I wanted to reach you or Jonathan."

"Jon's right here. I'll pass the phone; it's Preston Gates."

"Hello sir."

"Hope all is well with you Jon. I'll get right to it. An analyst manager at NSA filed a memo describing his contact with Karl and Karl's request that I be included. It created quite a stir and raised curious eyebrows. Did you know about it?"

"Yes."

"Are you able or better put willing to provide some background?"

"Yes and no."

"Typical Dr. Reynolds response," Gates chuckled.

"What I mean Mr. Gates is we're involved in this project and ran across some very interesting and potentially troubling communications that might fall into your sphere of interest. We're not finished. It might be a week or two – maybe longer. Karl's call to an old ACT contact was to see where might lie a department with a common interest should our initial concerns become real. As you are the only person back there we completely trust, it seemed logical to give you a heads up."

"Thanks for your confidence. Is your message: be patient and you may contact us, me again?"

"Yes."

"Thank you then. Best to you and Katie. Goodbye for now."

Jon hung up and shook his head, "The old guy is just as wily and smooth as ever. Maybe Karl's call was premature. I haven't come up with a pretext for calling any of those numbers, but I want to. My instincts tell me there's a very dangerous communications network out there."

"I see your point love. We don't speak the languages; we can't call and say 'wrong number' and if we screw up we could lose our Guzman lever. At the same time I'm bothered with our deal: our safety in exchange for allowing the drug trafficking to continue under law enforcement's nose."

"I know what you mean but the practical reality says we have no choice unless we enter some type of BS witness protection. No one is going to protect us adequately except ourselves. We all agreed to that."

Katie left to refill their glasses and upon her return burst out, "I just had a wild assed idea. I'll bet those twins we met at Christmas speak all those languages. Is it a ridiculous idea to involve them?"

Jon didn't respond immediately. Katie waited.

"Yes and no," Jon laughed at himself, "I'm so close to the problem maybe I'm not objective. I just have this desire to see what's behind the curtain. My nature, I guess. Why don't you run this by Michael – see what he thinks."

"I will. In the meantime how about some 'afternoon delight'?"

After looking at Katie in her bikini for several hours, he said, "What a great idea."

Chapter 12

Michael listened carefully to Katie's description of Jon's desires and language problem and Katie's off the wall idea, all without comment. He told her he would mull it over and be back in touch.

Instead of discussing this with Ruben Valenzuela over the phone, Michael opted for a day trip with prior notice. Now using his private car, he crossed the border and drove south to the Valenzuela estate where he was warmly greeted as usual. It was very quiet – no sign of children playing. Michael asked about that.

"In school, plus some field work is needed even in January. What brings you, I'm guessing not social."

"Astute as usual – I know you carry many secrets and have a perfect record of protecting those secrets. I'm going to add a few and seek your advice before a request is made."

"Exciting," the old patriarch said.

"You provided an equally astute warning at Christmas, however we have been living under that since returning from Washington DC last August. Shortly afterward we received a very direct death threat from someone we later confirmed to be Guzman or his organization. That threat included everyone on Katie's and Jon's team. I'm sure you read about those Seattle policemen and the attempt on me."

Michael went on to describe the various activities the team had undertaken, leaving out the location of their base of operations. He described Jon's plan and the Guzman phone call.

Valenzuela interrupted, "I would normally say he, they, you are playing with fire, but from what I know about Jon, I don't think so. What kind of genius is that man?"

Next Michael described without much technical detail, the communications monitoring system Jon and Karl had set up and the communications between Guzman and the Middle East including the use of a voice translator app. Michael concluded with an identification of the languages: Pashto and Dari, knowing full well Sr. Valenzuela would connect the dots.

They sat quietly for many minutes. Sr. Valenzuela showed no reaction but Michael knew he was several steps ahead and thinking as his eyes traveled about the room. In time Valenzuela broke the silence.

"What can you tell me about the danger? Those two are under both my guidance and protection."

"None that I can see. I didn't tell you before; we have a totally secured base of operations here in Mexico. That is where Jon and Katie are now. That is where the monitoring equipment is located. That is where their services would be needed. I can't see this would take much of their time but can't be certain of that. What other thoughts or questions might you have? I'm almost sorry to have brought this to your door."

"But not really, you old cop," Valenzuela smiled.

He went on, "They are back in classes now. The campus is about 500 kilometers south of here. A good road but still six hour drives each way. They have not been happy here since their parents' deaths and want to be in the United States, but because they are considered Mexican you know what that's like. As I think more about this, I don't want any of my family to have any knowledge of what you have described. Not even Ramon, their favored friend. I also believe it would be best if Jon and Kate met them on their own turf. If helping you would help them immigrate, there would be no question of their cooperation – just as long as it is safe. I will have it no other way, Miguel."

Michael's nervousness and apprehension melted away. He stood up and grasped Sr. Valenzuela's hand.

"I will think some more and communicate with you the old way. Would that be satisfactory?"

"More than acceptable my honored friend."

Jon was analyzing the latest call logs and database entries when he heard Katie on the phone downstairs. He guessed it was Michael. He was surprised at the length of the call. As he descended the wide staircase, she finished.

"That was Michael. We have a lot to talk about. It's about the cocktail hour so let's fix a drink and sit by the pool." Once seated, Katie began.

"Michael visited with Ruben Valenzuela at his home. I gave Michael the green light to disclose as much as needed."

Jon nodded his OK.

"Valenzuela was supportive but not enthusiastic and wanted no one in his family to have any knowledge. The twins are in school, near the coast about 600 kilometers south of Nogales and about the same from here. Valenzuela is willing to set up a meeting with conditions such as you and I must be there. But most important he needs to be convinced by us that it is not dangerous for the twins."

"How do we do that? We're walking targets and being around us is a risk."

"I don't think that's what he's concerned about, but we'll just have to see. By the way both Valenzuela and Michael will be at the meeting. They will drive down."

"Set it up please."

The Lear charter company was surprised to hear from Jon so soon. Jon gave them the desired pick up time at San Jose del Cabo and provided the destination to be Ciudad Obregón where there was an international airport.

Katie emailed Michael the specifics and shortly thereafter received his confirmation. There was a PS: we will pick you up at the airport. Sr. Valenzuela wants to meet first and has made arrangements, plan on overnight.

The Lear was waiting when they arrived at the airport. The total flight time was just over an hour and Michael could be seen leaning against a large white Suburban next to the terminal building after their touchdown and taxi. Jon pointed him out to the pilots as there were no airport personnel providing directions. Jon asked if they could make their own arrangements and both the pilot and copilot said yes. The copilot looked at his cell phone and announced he had coverage.

Sr. Valenzuela was behind the wheel. He waved and smiled, then indicated they should get in. Michael threw their overnight bag in the back and opened a rear door for Katie. It was a short drive into downtown Ciudad Obregón, a very clean and prosperous looking city. They pulled into the Gamma Fiesta Inn. Rooms had been booked for each. It was almost the lunch hour and Valenzuela suggested they meet once settled in.

The Inn Restaurant had several small private dining rooms. They occupied one and placed their lunch orders. It was understood Sr. Valenzuela would open the conversation.

They ate quietly, complimenting the food amid idle chit-chat. Once done Valenzuela stated, "I am very sorry to learn of the threats to your lives and how that must affect day to day living."

Katie responded, "Thank you. We have accepted this unfortunate consequence of our mutual activities last year. But we are determined to bring the threats to an end."

"Michael told me. I must confess to a great pleasure with El Chapo taking that call. He is a pig.

"More to the point you must understand my concerns about involving the twins. Their names are Ricardo and Lucinda, chosen when they arrived six years ago. They took my last name when their parents died but I have not adopted. Despite that I am their education sponsor and protector. My son Ramon is their best friend. I must know if there is any danger or chance of harm should they assist you?"

Jon decided to answer, "I cannot provide any guarantee if that is what you seek. All of our team has a target on our backs but we have taken every precaution to protect one another. What we intend to ask these young people to do will not place them in harm's way."

"Am I allowed to know the location of your base? Michael said it was in Mexico."

Katie answered, "A private villa behind two 24/7 security gates in the Palmilla area of Cabo."

"Don't you Americans call that the 'high rent district'?" Valenzuela chuckled.

"I guess it is."

"I am satisfied and will place my confidence in you. Ricardo and Lucinda will be here about four, after their last class. I am in Room 422. Please come there before. La Quenta, por favor, lunch is my pleasure."

Jon and Katie knocked at Room 422 a few minutes before the designated time. Michael opened and let them in. Valenzuela's room was a suite with a seating area for at least eight using the dining chairs. Jon and Katie sat on a small love seat. About fifteen minutes later another knock on the door was heard. Michael got up again and said, "Hola."

"Hola Sr. Ortiz, mucho gusto." It was Lucinda's voice. The twins entered dressed like American college students: jeans, colored T-shirts with logos and sayings and the requisite backpacks. They stood while Jon and Katie approached.

Sr. Valenzuela introduced them, "This is Dr. Jonathan Reynolds and Ms. Kate Ruiz. You may remember them from Christmas. Jon and Kate meet Ricardo and Lucinda Valenzuela, my friends and surrogate children."

Lucinda said in perfect, unaccented English, "It is my brother's and my pleasure, we know of you."

Jon was and was not surprised. They all sat down and Valenzuela picked up the introduction.

"Jon and Katie are involved in a project that requires translation assistance. Miguel is involved also. With several others, they work as a team. As the two of you are multi-lingual they are seeking your help. Two of the languages are …" He paused.

Jon answered, "Pashto and Dari."

A strange look came into the eyes of both twins but it wasn't fear, maybe apprehension.

Ricardo spoke for the first time, "Do you remember our parents?"

Jon felt unease with something nagging at him, "I don't think so but when I saw you at Christmas and we made eye contact several times, there was some kind of recognition that I couldn't pinpoint, sorry."

"Do not be sorry, Dr." Lucinda interjected, "It was many years ago in Seattle Washington at the University."

The light bulbs went off for Jon and he remembered, "The Applied Mathematics Seminars I did, right?"

"Yes," said Ricardo, "and we still have a copy of your doctoral thesis today."

Katie decided to lighten the mood, "I'll be darned, and Jon's sordid past finally catches up with him. Let me show you what I deal with day in and day out. As I recall when checking him out, his thesis was titled Complex Fourier Transforms for Digital Signal Processing. That's quite a mouthful but believe it or not it was damn near a best seller."

Everyone started to laugh; tension was broken. Michael reached over and squeezed Katie's hand and mouthed 'thanks'.

"Guilty as charged," Jon said meekly, "it is coming back. Was your mother's name Mahsooma and your father's Aasif? Did I pronounce that correctly?"

"Yes and near perfect," Lucinda answered, beaming. She had a lightening smile.

Ricardo added, "I am impressed you remember. It was 1995. The University in Kabul had an exchange program with the University of Washington. Our parents had just begun their tenure in the Mathematics Department and were at the top. We were almost eight and it seemed like a dream to travel to the USA for the school year. As we had learned English we spent … I think you call it fifth grade there. Our parents were in the graduate program and attended every one of your lectures and tutorials."

"I completely remember now. What an incredibly small world."

"Could you explain what you would like us to do, Dr. Reynolds?" Lucinda asked.

"Please call us Jon and Katie. We have a facility in Cabo that is monitoring low earth orbiting satellite communications. Recently we captured a large number of those in both the Dari and Pashto languages. We have a software based voice translator incapable of detecting the language nuances. That's first. Second we need someone to use our system and call back using a generic script to those communications. That someone must have a command of the language. Do you possess the language skills and have an interest? It can't be done from here so a short sabbatical or vacation is necessary and you would be compensated for your help."

Ricardo's eyes brightened, Lucinda looked somewhat wary. Jon thought she reminded him of Katie. Michael and Valenzuela continued as silent observers. And just like Katie, Lucinda wanted to know more.

"So we would have to relocate for a while to your Cabo place? How long might that be? Maybe we could complete some of our classes online. I don't understand about making those calls. Could you elaborate?"

Jon chose his words carefully, "Neither Katie or I are capable of a conversation in their language. Our information so far shows no knowledge of English on the receiving end. The subjects and scripts for the calls we would like you to make are not included in this conversation, but I will say they are not threatening, offensive or illegal."

Lucinda looked at Katie, who was nonplussed, and noticeably backed off.

Ricardo wanted to know when.

Jon said, "As soon as you could be available."

Valenzuela interjected, "I have been satisfied you will be safe and in good hands."

Lucinda looked over at her brother. Ricardo nodded.

"I guess we have a deal," she said, "We didn't discuss the pay."

God help me, just like Katie, Jon thought.

"Fifty US dollars per hour each plus all expenses."

"Wow," exclaimed Ricardo.

They said they could be ready to go late afternoon the next day. All shook hands and Jon and Katie went downstairs to the restaurant for a bite to eat.

"Who was that Katie, you're alter ego?"

"I can relate to her. What a looker, and Ricardo reminds me of Ricky Martin. They actually look Hispanic with their coloring. The only differences are in the eyes and nose shapes."

"I wasn't referring to that."

"I know."

Sr. Valenzuela and Michael first dropped off Jon and Katie at the Lear then departed to pick up the twins. Half an hour later the big SUV pulled in and Lucinda and Ricardo got out, fetched their bags from the back and looked around trying to figure out where to go. Michael pointed at the sleek jet and they saw Katie waving.

Ricardo used an American expression, "Are you shitting me?"

They said goodbyes to their mentor and Michael and together walked to the clam-shell jet entry stairs. Katie got their gear stowed and both were seated. Jon was typing something into his phone and just waved.

In less than five minutes they were airborne and heading due west.

The twins were awestruck and said nothing as they looked out the windows. They still hadn't uttered a word when they landed, loaded their bags into the Expedition and headed down the toll road towards San Jose del Cabo.

Passing through the security gates Lucinda finally spoke using another American expression, "So this is where the poor people live."

Katie snickered.

Upon their arrival at the Palmilla villa, Katie played host and organizer. "Assuming you want separate quarters, there's an apartment on the far right and guest bedroom upstairs, second door on the left."

"Who else is here?" Lucinda asked.

"For now it's only us. Michael, Miguel, and his wife Corinna use the far left apartment. Jon and I are next door. Other friends you haven't met use the near right apartment.

Upstairs is our office and a library beside the guest bedroom and bath. Wander around and take your pick."

They did and a short while later announced they would take the far right apartment together as there were two beds.

"Unpack and come out for a snack."

Katie had called ahead for another catered meal as there was no time to shop or cook. It arrived and was spread out on the tables by the pool.

"Come and partake," Katie shouted as she saw them exiting the private door of their apartment.

Jon joined them a short time later and without another word began to devour dinner.

Katie asked, "I don't mean to offend but would you like something to drink?"

Ricardo laughed out loud, "We may have Islam in our veins but screw that, we do have alcohol now and then."

"I'll get some glasses; the Sangria is outstanding."

With dinner over it was getting dark so after cleanup they sat in the villa great room in uncomfortable silence.

Katie asked, "Is something bothering the two of you? Please don't be bashful around here. We need to have very open and candid lines of communication."

"I don't know about Ricardo here," patting his knee, "but I think we're almost overwhelmed and it has happened so fast. While we have seen and been in awe of our mentor's family and his property, this is several steps up. Only TV has shown us this kind of life and we have accepted it's not obtainable for us."

Katie reached out her hand for Lucinda and said, "Be my friend; let's take a walk on the beach; I have many things to share with you."

They left and Jon knew Katie would share her history, life experiences and hopes for the future. What an incredible woman.

"What are you smiling about Dr. Rey ... Jon?"

"How lucky I am to have found such an incredible woman – no, a soul mate – as Katie. I'm not whole without her," Jon couldn't figure why he had said that to a near stranger.

"Both Lucinda and I have read her stories. Almost everyone knows who you are."

"That's part of the problem."

Katie and Lucinda returned and announced everyone needed a good night's sleep.

The next morning Katie was whisking some eggs when Lucinda appeared looking awake and refreshed.

"Can I help you?"

"Thanks, how about toasting those muffins while I scramble these."

The smell of coffee awakened Jon and Ricardo who appeared shortly thereafter. They filled their cups and went out beside the pool promising any help needed.

After breakfast Katie asked if the twins felt comfortable enough to share how they had arrived in Mexico.

It seemed the barriers had been broken down overnight. Taking turns Lucinda and Ricardo shared a remarkable story.

Their parents were the result of an arranged marriage but soon loved each other. Their ancestors had come from the northwest area of Afghanistan beneath the Hindu Kush in the village of Skazar and despite the subservient position of women, both parents had been accepted to study at the University of Kabul. They were brilliant students, both having an affinity for languages, but mathematics is what interested them the most. The twins were born while the parents were still in university in 1987.

Both parents had kept a low profile because of the Russian occupation and while the war raged with the Mujahedeen. The Russians left in 1989 and a corrupt government emerged and took control. As a female, Lucinda was fortunate to be a baby but her mother suffered greatly. She was tolerated for a short while because of her exceptional intellect and reputation at the university. They led a fearful existence for the next ten years. Their only break had come when their parents had been selected for the university exchange program. By then the Taliban were in control. They did not want to return but alternatives were not available. Upon their return, their mother was removed from her teaching position and made to direct a group of similar aged women in religious training and indoctrination. All were required to conform to the requirements and beliefs of Sunni Islam.

That group, at least on the surface, performed as directed, because more than one of the members had been removed and publically executed. Their father continued to teach. When 9/11 occurred and there was much rejoicing in the streets of Kabul by the Taliban, their parents prepared to leave. They took an unauthorized trip to their home village under the Hindu Kush, leaving everything behind. Two months later they saw their opportunity and along with many others, made the arduous journey through the passes to Peshawar across the border in Pakistan. Their father traveled to Islamabad and sought asylum at the US Consulate. That was not possible because there was no way to verify the family or their loyalties. After failing again with Canadian authorities, Mexico accepted them.

"And so here we are," stated Lucinda. "Our memories are very sharp from about four years old."

"And I thought I had a tough upbringing," observed Katie, "I don't have clue one."

After listening, Jon's feelings ran from more anger to empathy to determination to do his part to right things if possible.

Chapter 13

The next morning Jon was in the upstairs office sipping on a cup of coffee and reviewing the captured communications and the text translations since they'd been gone when he sensed a presence behind him. He turned and greeted Lucinda and Ricardo at the open door.

"Is this your control and command center," Ricardo asked.

"Something like that," Jon responded, "good morning to you. Did you sleep well?"

"Very well," Lucinda responded, "what are you doing?"

"This is the backbone of our transmission capture system. I'll show you more later. Let's get some breakfast. I hear Katie down there." He got up and they followed.

Seated at the table cutting their French toast, Jon asked, "How did you acquire your language skills?"

Lucinda answered, "Our parents had an affinity and ear for languages. Between Ricardo and me we read, write and speak seven. The Pashto and Dari are obvious but our parents insisted we learn Arabic. We began to learn English when we knew we were going to Seattle and after a year there, we were pretty fluent. Spanish came later. French was a requirement in school here."

"Impressive," commented Katie.

"When do we go to work?" asked Ricardo.

"After breakfast," said Jon.

They returned to the upstairs office, pulled in two extra chairs, and listened to Jon's instructions.

"We are going to place two calls, both from this, a satellite phone. The numbers one and two are on this card. Also the card shows the language: one is Dari, two is Pashto. Card three shows the script and there is a blank I hope you can fill in. Please read the script first." It said:

'Hello, I am seeking .... is he available?'

"The answer will be 'No' so say thank you and hang up.

"To make the call real or at least believable, please edit my script to conform to what someone in Afghanistan would expect. Also the blank represents someone important in the current government you can think of. It must be male. I also think Ricardo should make the calls for now – a male's voice."

Lucinda and Ricardo took the cards and pondered Jon's comments.

"I agree my brother should make the calls. A woman's voice would be immediately suspicious. Also I recommend you ask for Hafizullah Amin, a former Prime Minister now dead, but his son of the same name lives and is important. Do you agree brother?"

"Yes, I can do this. Tell me when."

Jon handed Ricardo the phone and showed the few buttons and icons.

Ricardo made the first call in Dari. It was answered and after a very short exchange terminated. Ricardo was stressed so a break was taken. The twins looked in fascination as the call verbiage appeared on the screen.

"That's what I said in English," Ricardo exclaimed.

He was calm enough to make the second call in Pashto. It went a little longer but ended the same.

The text of that call appeared shortly thereafter.

"And what does that prove?" Lucinda asked.

Katie said, "Watch."

Jon brought up Google Earth, invoked the GPS option, pasted in the coordinates of the first call and clicked search. They watched in fascination as the globe turned and zoomed in on the coordinates.

Lucinda let out a screech, "Oh no, that's where Omar is supposed to be located."

They all knew who Omar was.

Jon repeated the process with the coordinates of the second call. The twin's response was more muted this time.

"That's almost in the center of the main poppy fields. I don't remember the name of the village."

It was quiet for many minutes. Katie and Jon left the twins and walked out to the pool and sat.

"Do you think we showed them too much?" Jon asked.

"I don't know; let's see what happens."

They noticed the lights come on in the twin's apartment almost an hour later. Jon and Katie were still by the pool, the twins approached.

"We love American expressions. So what's the rest of the story?"

Katie laughed. Jon shook his head and said, "You two are amazing, welcome to the team."

Lucinda and Ricardo sat down.

The next morning there were no further discussions or explanations of the night before. At Katie's urging Lucinda and Ricardo were in the surf splashing each other. To relieve the tensions Katie arranged a fun day taking the twins shopping and to explore Cabo San Lucas leaving Jon behind to his own devices. As they pulled out in the Expedition Lucinda asked, "Are you going to tell us what's really going on and what was that 'team' comment last night?"

Katie answered, "Jon may have spoken out of turn but I doubt it. He almost never does anything without a purpose. I can't predict what our group of friends will decide, but I for one could not be thankful enough for the two of you. An entire new perspective opened last night."

When they returned late Jon had not left the office other than to make some physical adjustments to the digital satellite receiver hidden in the sea side yard. He was frustrated. No amount of adjustment or tweaking would allow him to connect to low orbit satellites beyond the east or west horizons. He never heard them return and was deep in thought when Katie stuck her head in the door and told him a pitcher of perfect margaritas was waiting by the pool.

He needed the interruption and joined them poolside.

"Tell me about your day at the office dear."

Jon grinned and growled. The others laughed.

"Truthfully I could have taken our receiver and pitched it into the Sea of Cortez."

"What is the problem lover?"

"Connecting to those god damn low orbit satellites that require line of sight."

Addressing the twins, "Do you know anything about the various islands west of India in the Arabian Sea?"

"Oh Jesus, here we go again," exclaimed Katie.

"Nothing," both replied.

"Thank you God," Katie added, "what is up with you Jon? I really like it here."

"We need a group meeting. The MLK holiday weekend is coming up. Would you contact everyone and set it up, then call the Lear charter company for transport? I would really appreciate it. Until then I'll keep my mouth shut, however two more calls occurred today that Ricardo needs to call back."

"Whatever you say love, but patient I'm not."

Karl and Marie pulled Stephanie out of school for Friday and were picked up late Thursday. Corinna and Michael accepted an offer on their house, but were confident they could close and vacate by the Lear's arrival time. Katie and the twins became better acquainted and stayed out of Jon's brooding ways.

After everyone had arrived and the twins introduced, instead of receding as Jon had witnessed at Christmas, they became outgoing and effusive. For the first time in days Jon sat back viewing his friends and began to relax. Katie came over, sensing what he was feeling and wrapped her arms around his neck and nuzzled.

"What have we done? What have we created?" Jon asked softly.

"I don't need to answer that question; you already know the answers – be thankful. I love you."

The competition to be helpful in the kitchen for breakfast was a sight to behold. The men, including Ricardo, sat poolside on their best behaviors waiting for either an order or food delivery.

As it turned out a delicious breakfast of muffins, bacon and chorizo scrambled eggs were served. Afterward Katie stood and clicked her OJ glass several times for attention.

"Thanks for coming. It's great to have you here again. Jon and I will do clean up. Enjoy the pool, beach and sea. This afternoon we will lay out the issues and problems we face and make some decisions."

As Karl was not much of a sun worshiper, he helped with cleanup then joined Jon upstairs for a briefing.

Karl came away with many conflicting thoughts.

Jon reached out a hand to Katie in the pool, "We should gather the group and lay it all out."

Katie agreed and got out to dry herself and assemble the group.

They were seated around the pool area.

"First," Jon began, "I want you to know I briefed our new friends Lucinda and Ricardo about us – where we've been and how we decide where we're going. And I do mean decisions. That's why we're here."

"I'm moderately confident Guzman got the message. There have been no further calls indicating a threat against any of us. But we can't be certain yet – continued vigilance please. The larger issue and the reason I asked you to return is borne in my conscience.

"We spent the better part of last year destroying or at least disrupting one drug cartel after the other. I'll let Katie speak for herself but my initial motive was revenge for the deaths of my wife and children. Of course it grew beyond that. We all know the unintended consequence was we became targets of those we damaged. This Guzman jerk is at the front – hence our presence here. While conceiving and implementing a plan to deal with Guzman we stumbled into another world.

"We knew drug trafficking is a worldwide business not simply confined to Mexico and a few South American countries. We also know the US is the largest market, consumer of these various drugs. Satisfying that demand is where the Mexican Cartels fit in. Afghanistan is one of, if not the largest producer of heroin and its raw material opium. We all know what's happening there right now. In the course of monitoring Guzman's satellite phone calls we discovered he has a pipeline into Afghanistan through distributors in Pakistan.

"I guess that shouldn't be surprising as the Asian producers want access to the US market and Guzman provides that. What is surprising, at least to me, is who is at the head of the heroin chain: the Taliban and Al Qaeda, the same terrorists responsible for 9/11.

"They're financing their attacks and the ongoing wars using drug sales. That's my belief. Lucinda and Ricardo all but confirmed it. The question is what can we or anyone do about it? Or should we do anything? I want to hear your thoughts."

Katie added, "Consider the insanity of it all. US users of heroin are paying for a war against us."

Michael was astounded, "Are you telling me your system or process connected through Guzman to Al Qaeda? And you were able to pinpoint their location?"

Jon nodded.

Michael continued, "The intel value of that is beyond belief."

"I know."

Dripping with sarcasm, Karl said, "Here we go again. Maybe there's a resort for sale in Kabul that we can buy and set up shop. But I don't think they have any beaches."

That drew a few chuckles.

"Seriously," Karl went on, "We're on the outside looking in. It's almost like what we did last year: finding the head of the money laundering network, only this time it's a communication network. Think what could happen if we recovered one of their phones and tracked all the calls, cell and landline included."

"I thought that's what NSA does all the time," Marie added.

Jon responded, "They may try with wire taps and capture from a few cell towers, but they don't know or have a starting point like we did with a captured phone and actual numbers. They can only focus on selected geographical areas and would be sifting through billions of calls looking for suspicious verbiage. And should they obtain a phone, probably a cell, it would be an untraceable throw away. In all probability only the leaders would have satellite phones.

"And on the subject of location or selected geography, I've become very frustrated trying to expand our capture range. It's just not possible with the receiver technology we're using. We or someone would need to have hundreds of our type of ground systems in place, after first knowing where to place them and what to look for."

Karl added, "And they don't have Jonathon Reynolds to decrypt the calls."

Corinna asked, "What can we do? We're only six people? I do see the insanity but it feels almost overwhelming."

Marie spoke up, "I do see your point, but to do nothing, ignore what we know seems wrong. The first thing that flashed through my head as I was listening was the image of those planes crashing into the Twin Towers."

Jon added, "That's what I keep coming back to. It's why I said 'conscience'. Our options are clear: one, assume we have Guzman in check while continuing to monitor his communications and trying to conduct normal lives – all the while turning a blind eye to what is really occurring in the drug world, or two, become proactive again. I'm really torn, my friends."

Corinna asked another question, "Jon, can some national security department use your system?"

Karl answered first, "Yes it's possible. The drawback to our system, as Jon said, is it requires line of sight to a particular satellite. So when someone makes a call from Kabul, for instance, and depending on the time of day, the particular satellite routing the call would have to be within our receivers' range, horizon to horizon. I suppose it could be operated from some Army base in Afghanistan, but they would need to know ahead of time what number or numbers to capture or simply capture all and decrypt. That would take an entire ground station like the providers have. And it's certainly not mobile. Our fortunate situation here is Guzman's Sinaloa operations use the same satellites we're tracking. If he moves elsewhere we're out of luck. Does that make sense?"

Katie added, "And we all know the issues and pitfalls of dealing with any government agency. God help us and my conscience is just as torn as Jon's."

Jon then said, "Regardless of whatever option we choose, we must keep in mind the primary reason for this place – our safety. Katie and I don't view this place as temporary. We think it is a terrific vacation spot for all of us to share over many years. But we're not out of the woods yet, so the monitoring must continue. For how long I really don't know, several months anyway."

Many minutes passed in silence.

Corinna broke the silence, "I am proud to be included on this team, but also know I'm a supporting player, at best. Being Michael's wife and partner all these years, I've never lost sight of what those words on his squad car mean: 'Protect and Serve.' Selfishly, I would love to tune all this out and try to enjoy our retirement, but as Jon said: it's about our conscience and who we are as individuals. I'm ready to support Jon's option two."

During another prolonged silence, Michael reached out and took his wife's hand and held it tightly. Marie and Karl locked eyes, smiled, and nodded to each other.

Marie said, "We're on board."

Tears welled up in Katie's eyes.

The enormity of it all was not lost on anyone.

With a renewed determination, Jon asked, "Michael, Corinna, would you consider living here for a while? And Karl would you set up a VPN from your home to the system here?"

Michael said, "Thought you'd never ask."

Karl said, "Of course."

"That way," Jon continued, "Michael and Corinna are safe, the system keeps running and Karl and Marie can support the monitoring and technical assistance from Seattle."

Both couples agreed they could do that.

"Katie and I are going to renew an acquaintance, AG Preston Gates."

"Dr. … Jon," Lucinda spoke for the first time, "Ricardo and I want to help if we can." Her eyes swept across all the faces. "When we were introduced, you were not told we were born Northwest of Kabul and immigrated to Mexico after the war started five years ago."

Ricardo added, "That is true and we have technical skills in addition to the languages. Please let us help."

Chapter 14

Corinna said it would take less than a week to move to Palmilla and about the same amount of time for Karl to install and test the VPN (Virtual Private Network). The Lear returned the Ortiz's and Oberling's to Tucson and Seattle, respectively. The twins stayed behind. As Michael boarded the plane he told Jon quietly to remember their promise to Sr. Valenzuela. Jon nodded affirmatively.

Fiber optic cable had been laid when the Palmilla Villas were being developed so it was quick and simple to bring in dedicated Internet service. Karl would use that for the VPN. Jon purchased Lucinda and Ricardo a laptop each and while they were catching up on their classes, he and Katie discussed the twins' offer of assistance at length. Both were sorely tempted to include them but in what capacity they weren't sure.

Two days before Michael and Corinna were scheduled to return, Karl had the VPN up and running. He and Jon thoroughly tested both ends of the pipe. That same day Lucinda asked during dinner if their request had been considered. Jon said yes and explained the promise he had made to Sr. Valenzuela.

"I'm between a rock and hard place."

Ricardo retorted, "I think we should have something to say about that."

"You're probably right."

Katie spoke up, "I have an idea. After Michael and Corinna return Jon and I are going to Washington DC. On the way we will take you back to Ciudad Obregón. We expect to return with some resolution to the terrorist connection. Should that happen we would meet with you and Sr. Valenzuela either there or at his estate and see where it goes. Does that work?"

Lucinda and Ricardo brightened considerably.

A dark blue government limo was waiting when the Lear landed. It was deja vu as they had landed at the same place less than six months earlier. They were taken to the same base house used before. They were told Attorney General Gates had invited them for dinner and could they be ready at 5 PM to be picked up.

The reunion was very friendly and comfortable. Katie had chosen to dress down for the occasion in her usual black pant suit with a white shirt and high collar. Jon had dug out his best sport coat but no tie. AG Gates loved Katie's new hairstyle. They played catch-up, with Gates explaining his task force was still up to its ears in support of prosecutions.

"However, I'm very sorry to report, Katie, the President's Blue Ribbon Panel that was to make recommendations curbing the demand for drugs has accomplished next to nothing."

Katie responded that she knew that and was very disappointed but not too surprised.

"So Jon, the two of you were circumspect at best about the reason for your visit. Care to elaborate?"

"Sorry, not here and not tonight. We request you put together a small trusted group to hear our situation. We would like to include Karl's contact Randy from NSA, perhaps the National Security advisor or someone from there and lastly someone from the Pentagon in the intelligence area knowledgeable about Afghanistan, and yourself of course. Also, no notes or recordings, at least for now."

Gates appeared almost shocked but recovered quickly, "Can I at least know the topic?"

"More than one topic, but how about the US is financing terrorism and the enemy in Afghanistan via Mexico."

"Holy shit, oh pardon me. I'll see what I can do. Should Homeland Security be included?"

"Your choice, I don't care as long as there are no turf wars. Katie and I are available anytime and will wait until you can set this up."

"I do have one other topic, if you don't mind."

"Go right ahead."

"Speaking of Homeland Security, I believe that INS was moved from Justice to them."

"You are right."

"Katie and I are thinking of sponsoring nineteen year old twins who want to immigrate from Mexico but have not been successful for a variety of bureaucratic reasons. They do have some baggage through no fault of their own. Do you have any contacts who we may speak with?"

"Of course, we're just one big happy family in DC," he joked. "Let me work on it and I'll call you in the morning about the first request."

They left and were driven back to the base house.

Once there Jon asked, "How do you think it went?"

"Thought you gave the old man a heart attack for a moment, otherwise a nice evening."

Preston Gates called before nine. He said the meeting was set for 1 PM at Justice, third floor conference. They would be picked up at 12:15 because of traffic. Jon confirmed.

They got off the elevator and were escorted to the conference room. They were introduced all around. Jon had never met Randy before. His last name was Allison. Seated on either side of AG Gates were Clinton Prescott, the National Security Advisor, and a Lt. Colonel in full Army uniform. Another was seated next to Randy and she was introduced as Assistant Director at Homeland Security for counter-terrorism. Her name was Shelia something.

Gates began with, "Needless to say everyone here is well acquainted with you two. You have our admiration and gratitude. Dr. Reynolds?"

"Thank you all for being here. I trust AG Gates has informed about no notes or recordings at this time. The reason will be obvious in a few moments. Katie will provide some background first beginning with an email she received in September, Katie?"

For the next ten minutes she led them through the threats, the Seattle police killings, the attempt on Michael Ortiz and ended with the capture of the shooter's satellite phone. Jon took it from there.

"There was a very short window when that phone was connected to the Globalstar network and we obtained three phone numbers. Our local DA, Wes Davis, was able to subpoena the phone records for two as the area code was 206. One was a throwaway Trac phone, the other belonged to a POI suspected to be the leader of a local gang. We determined the other number called was a satellite phone belonging to one of Guzman's lieutenants. How we did it we'll get to later but I want to take you to the bottom line quickly.

"From that number we obtained the SAT phone number for Guzman himself. I spoke to him or should I say he listened. Further we know the exact location where he took that call. Now it gets real interesting.

"We intercepted several other calls with multiple exchanges between them. One was between Guzman and Zambada, aka El Mayo, another was to a heroin distributor in Karachi Pakistan, and several were made to two locations in Afghanistan. Are you familiar with Gostan valley, Nimruz Province and Arhhandab?"

The Colonel almost fell out of his seat. Gostan valley is one of the largest poppy growing regions, second only to Helmand. They are in the control of Al Qaeda; Arhhandab is the site of Mullah Omar's former compound. Everyone started talking at once. Gates rapped his water glass on the table until calm and quiet prevailed.

Katie spoke very slowly, "So now you see the insanity. A US citizen buys and shoots up heroin. The price he paid finds its way to Guzman in Mexico. Guzman buys the heroin from terrorists killing our soldiers."

Not a word of response came out. Everyone was staring and thinking.

Randy was first, "Where did you come from Dr. Reynolds, another planet?"

Katie responded, "There have been times when I've wondered the same thing."

That broke the tension; Jon was nonplussed.

Shelia stepped in, "Are you going to share your methods?"

"You know what Advanced Cyber Technology, ACT, my former employer did, does for a living. When Hughes was manufacturing the communication satellites, both hardware and systems, I wrote the encryption algorithm used for voice and data transmission."

"Did you provide that to us?"

"You never asked."

She started to speak again but thought better of it.

"Is there a story behind this Katie?" Gates asked.

"No, the outrage in the press and public that would generate could compromise national security and our troops."

"Thank you Ms. Ruiz," Prescott said.

"I'm at a loss for words," Gates offered.

"Never thought I would see that day, Preston," Prescott came back.

Katie wanted some breathing room and closure of the meeting before they would be forced to reveal anything more.

She said, "We are not going to create any plan or course of action forward today, in this room. Jon and I respectfully request each of you give our information your own analysis and formulate any questions. We will remain available in Washington until a conclusion is reached and decisions are made. I am comfortable with this group to continue meeting and sharing. If Mr. Gates approves it, feel free to add additional expertise, but absolutely no committees. Does that work for all of you?"

"I think it does," Gates' hand swept around the table. Everyone nodded or said yes.

Preston Gates was summoned to the Oval Office. Upon his arrival Clinton Prescott, the National Security Advisor, was seated along with the Directors of the FBI and CIA.

The President opened with, "What is wrong with us? What is wrong with our intelligence gathering that costs many billions of dollars? We might as well hire Reynolds and Ruiz for a million or so a year each and get rid of all of you. What say?"

"We don't know how he did it," the FBI Director offered.

The President was angry and just getting warmed up, "I don't give a god damn how he did it. Fact is he did. Is he that much smarter than our whole apparatus? Or is this administration infested with idiots?

"Thank God Katie Ruiz chose not to publish. Thank God Preston has their confidence. If there is one leak it will come from in here and I'll have your asses. Is that clear?"

They responded, "Yes, sir," in unison.

Preston Gates spoke up, "Knowing them as I do, there will be no disclosure or sharing of the technology associated with their discoveries until a plan is in place. That is their nature. But all of us better believe they are loyal citizens, even patriots maybe, and can be trusted to the nth degree."

"Thank God for that," the President said.

Gates went on, "Katie invited us to include other expertise. That DIA Colonel never said a word and seemed almost shell shocked. Is he the best we have? And who else should we call in, keeping the group small and the information close?"

The President answered the questions, "Brief the Defense Secretary and the Joint Chiefs then answer that question. I want the best and most appropriate person from CIA and FBI to join the group, but that's it. Briefings only through Clinton here, and remember about leaks, understood?"

The meeting ended.

Jon checked in with both Michael and Karl and learned that everything was running smoothly. Michael's sometimes acerbic manner showed when he reported all the recent captured calls were just the usual drug transportation instructions.

"Jon, lover boy, do you have any idea where we're going on this?"

"No real clue but a lot of desires and ideas."

"Care to share?"

"I only know that we must pick and choose what we can control. To do otherwise would suck us into the morass that runs this country, that suffocates rational decision making."

"I'm glad to not be that cynical. I remain optimistic. We're a good balance you know. Lay down with me."

Chapter 15

The next meeting took place the day after the President issued his orders. It was in the same conference room at Justice.

Jon and Katie were introduced to James Ferguson from the FBI and another James, who wanted to be called Jim, from CIA. The Pentagon Colonel had been replaced by a one star general from DIA named Gus.

As before Preston Gates took charge, "We're here to determine how and to what degree the information Jon and Katie imparted can be effectively used and expand on that. Am I right?"

There appeared to be general agreement.

Gates continued, "Jon? Would you field the questions we have?"

"I'll try."

Shelia from Homeland Security went first. She had a small notepad which Jon chose to ignore.

"I spoke with Randy after our last meeting. I read the request your partner at ACT made and reviewed the information Randy provided. Let me see if I have this right. You are able to receive communication signals from low earth orbiting satellites, somehow filter them and then decrypt the voice messages within."

"Very accurate description," Jon answered.

Randy interjected somewhat defensively, "We have several listening posts that capture similar transmissions but as the capture requires line of sight, it's limited and short term and we only obtain phone numbers called and received."

The one star said, "Dr. Reynolds, would you please share some details? I believe we're all guessing here and to move forward we need to have a level playing field."

Jon chuckled; he instantly liked the man.

"The name's Jon and I will."

Jon continued to explain what is required is a digital satellite receiver that can tune to specific frequencies. The captured phone was a Globalstar so they knew the frequency range. Next, they had the shooter's phone and a specific number. It was a no brainer to strip off the Globalstar front end digits and extract the likely phone number prefix. He explained how Sgt. Ortiz had obtained the locations for the identified prefix and that led then to Mazatlán, Mexico.

"So now we had the filtration criteria. We recorded any call with that prefix until we discovered a message of interest, or better said, importance to our personal safety. After that it was simply a matter of monitoring calls to and from that number. Every SAT phone has a built-in GPS for emergency purposes, so once a call is in process a location is easy."

"You didn't explain the decrypting procedure," Shelia stated.

"No I didn't."

"And?"

"No answer at the present time."

She was clearly frustrated.

The new DIA general spoke up again, "I'm sure that was a general overview and there's a lot more technical detail."

Randy asked next, "Because these satellite calls require line of sight transmission, I assume you are set up somewhere near Mazatlán."

"You assume correctly."

The CIA Jim asked how they had determined the location of the calls with Pakistan and Afghanistan.

"We called those numbers with GPS Tracker running on both ends."

"No kidding."

"No kidding."

Preston Gates took over, "Let's take this piece by piece and start with the Guzman connection. He was one of the few Cartel leaders not captured."

Shelia asked, "Shouldn't DEA be represented here?"

Katie decided to answer the question, "No and they're not welcome unless Mr. Gates insists. If you will forgive my crudity, they haven't learned a damn thing over the past six months. Why the President doesn't fire all their arrogant asses or send them on border patrol so they could learn something, I can't understand. Maybe they're so stupid they're unable to recognize a pink slip."

"Maybe I shouldn't be here," Shelia retorted.

Katie looked her squarely in the eye, "Play or pass."

Gates decided it was time for a break but asked Clinton Prescott to remain behind.

On the way out Jon gave Katie's backside a couple of pats; she smiled.

Behind the conference room doors Gates said to Prescott, "If that wasn't a classic example of our government in action, I don't know what is. How have we become so screwed up?"

"Preston, I live with it every day of the week."

"I actually enjoyed that HS bitch getting hers."

"Me too, but how do we move forward? Those two and the folks behind them could prove invaluable. After all we do have a job and responsibility."

"Let me try another tack when we resume."

The meeting resumed.

AG Gates opened again, "Jon, Katie, I had no idea what you wanted or expected when you contacted me. You didn't say. If it was a rapid, consensus supported course of action that isn't happening. No one here and all the others at our comparable levels couldn't make a unilateral decision if our lives depended on it. That's the way our system works and was designed. It works well most of the time. Even the President is indecisive and assembles a collection of opinions before making a decision, which unfortunately can sometimes be too little or too late. I know you two well enough to suspect you're already several steps ahead. What do you want? How can we cooperate? What support can we provide?"

Jon answered, "Allow Katie and I to discuss and consult with our team tomorrow morning. We can meet again in the afternoon at the same time."

Gates looked around the table and said, "That will work."

Shelia added, "I will be here too."

On the way out Gates handed Jon a card. It contained the INS contact he had requested. He is expecting your call, Gates informed.

Part 4 – Tentative Cooperation

Chapter 16

Jon set up a conference call using a free service. Michael and Corinna called in from Cabo, Marie from home and Karl from his office at ACT. They were brought up to date.

Katie told them, "Jon and I discussed the situation until very late last night. To repeat the obvious we're here because we want to do our part to help, motivated in part by 9/11 and our military fighting two wars. Jon, tell them our proposal."

"We will provide replicas of our system in Cabo, one primary, one secondary or backup. Katie and I will spend the better part of a month in Afghanistan on some strategically located base operating the system and training others. That base must have Internet service. Whether we ever go off the base remains to be seen. It should be safe enough. We will use those two phone numbers as the starting point and then see where that leads if anywhere. We will turn over the systems then return home satisfied we did the best we could.

"Oh and Michael, we're going to approach Valenzuela about using the twins. Their language skills would be invaluable."

"I guessed you would do that."

Marie said, "I'm not sure I could be that brave. My thoughts and prayers will be with you all the way, but I believe it's the right thing to do."

Corinna added her agreement.

"We're meeting again this afternoon. Karl, we'll call you to put the systems together if we reach an agreement. Bye for now and many thanks. You people are unbelievable."

They were all seated in the same chairs. Jon opened the meeting in a very direct tone of voice, "Let's cut right to the chase. You may turn on any recording devices this room might have. Katie?"

"From my writings you probably know I'm a hardened idealist. That's an oxymoron if ever there was one. Jon is an even tougher cynic but he has a deep conscience. Together we blend. Also from my writings and our activities last year, we hold some unfavorable views of this government. Today that is set aside. We remember 9/11 and respect the sacrifices our troops are making. We're here to help if we can.

"Until this moment we were driven by a desire to lead a safe, private life together and with our friends. That has not been possible. Motivated by the threats, the murders of our Seattle police friends, the attempted murder of Sgt. Ortiz, we undertook a number of actions. Please don't insult us with any statement that authorities can provide protection.

"To our utter surprise we uncovered the connection between the Sinaloa Cartel and possible locations of Mullah Omar and Al Qaeda in Afghanistan. This discovery is so overwhelming we decided to bring an offer of assistance to you. We have a plan. Jon will lay it out. It is minimally negotiable."

Jon allowed the others several moments for them to absorb Katie's words. There were no comments or questions.

"Our group has developed a hardware and software system that is capable of receiving voice and data transmissions from low earth orbiting communication satellites. Our focus has been on the Globalstar network. Beyond the simple capture, the system is able to apply filtration criteria for specific areas of interest, decrypt the communication and with some human assistance translate into English. The calls of interest are stored in their entirety in a summary database. The system procedures run automatically but require human monitoring and tuning. We are willing to provide this system for your use, conditionally."

Before Jon could continue, CIA Jim interrupted with, "What conditions?"

Jon paused then looked directly at him, "I'm about to get there.

"We will field your questions about the system, but first please listen to our plan for deployment and use. As previously mentioned the system requires monitoring and tuning. It is not turn-key at this time. I am the author of most of the software. Here is what we propose.

"We will recreate two of these systems, one primary and one for backup reserve. Along with these systems Katie and I will be flown to Afghanistan, along with possibly two others, and be set up at some base strategically located. We would work with Randy and whoever, to choose where based on the locations of our captured calls and the satellite orbits. We would spend up to a month there operating the system and training your personnel. Should the system prove useful and uncover the desired intelligence, we would return home leaving full control in your hands. I'm finished. We'll now field any questions."

There were a number of under the breath mumblings and some shifting of the seats. National Security Director Prescott spoke, "Simply amazing, just amazing. Preston you were right. Is there a precedent for all this? I never heard of anything like it before."

To everyone's surprise Shelia interjected, "Every one of us has been guilty of turf wars and sometimes the intelligence being circulated has been edited for a variety of reasons I won't go into. Beyond my thankfulness for citizens like Dr. Reynolds and Ms. Ruiz, I see a real opportunity for cooperation. Mr. Gates, do you know of any reason we can't proceed?"

AG Gates looked at Prescott, who shook his head no, and said, "None whatsoever and I want to echo her comments."

The buzz around the table began. Jon and Katie sat back and watched.

Shelia got their attention again. What a tough old broad, Katie thought.

"I have a practical question to ask you Mr. Gates as the DEA is not represented. What about El Chapo?"

Jon answered, "Our original system remains in place and continues to operate. Its primary purpose is to ensure no imminent threats against us. It now has a second purpose, capture any new Afghan or Pakistan phone communications and numbers. Katie, would you pass around the transcript of my calls to Guzman. Pardon the language."

Katie placed several copies on the table to share. Gates could hardly suppress a grin after reading. Others shook their heads. One could be heard saying, 'all right'.

Jon added, "I'm sorry to say you will have to catch him on your own. We will provide the Mazatlán location but that's it. The GPS coordinates are on the backside of that transcript. I hope you understand we're sick and tired of hiding out."

Clinton Prescott addressed the room, "Preston and I need to brief the President. Should a green light be given this is the control group. General you should conduct your own briefings and determine who might be point on this in Afghanistan."

After the meeting Jon used his cell phone to call the INS contact. They set up a meeting for 8:30 the next morning. Directions were given. The winter day was cold and blustery but Katie wanted to play tourist and visit the Supreme Court building. They were dropped there and told the driver they would get a taxi to the base entrance.

Early the next morning they called another cab that took them to the INS office. They were greeted warmly as Katie and her journalism were well known.

Katie presented their case describing in detail where Lucinda and Ricardo were born, how they arrived in Mexico after failing to enter the US. She described the deaths of their parents because of an accident, and the role of Sr. Valenzuela and his family. She went on to explain they took Mexican names including Valenzuela. They did not have birth certificates, only the Mexican entry visas from their parents. They have a strong desire to immigrate here and she and Jon would like to help and or sponsor. AG Gates could provide references if that would help.

The INS officer seemed genuinely sympathetic. He called for his assistant to bring the necessary refugee applications.

He said, "Please have them complete these. That's the first step. We need originals because of the signature pages. A cover letter from each of you and something from AG Gates supporting their immigration would go a long way. Please send the completed materials directly to my attention. It is a great pleasure to meet both of you."

They were standing on the sidewalk trying to decide where they should go when Katie's cell phone rang. It was Preston Gates.

"Good morning, The President has given your plan the green light. He also sends his best regards and a large thank you. We would like to convene again this afternoon. Would that work?"

"Yes, see you there."

There were two additions to the group, both in uniform. Only one chair was vacant in the large conference room.

The General spoke first, "Dr. Reynolds, Ms. Ruiz I would like to introduce Major Don Taylor and Lt. Charlie Strand. They work for the Defense Intelligence Agency. Both have done tours in Afghanistan. Don has a Master's degree in computer science from Georgia Tech and Charlie a mathematics degree from Columbia. We call them our whiz kids." The General smiled.

"A pleasure," both said.

Gates added, "While General Irwin will continue to oversee all military aspects of this operation, the Major and Lt. will be on the ground with you. I understand there are a number of questions. Let's get started."

Randy read from a list on a notepad in front of him, "What do we need to do for the two systems as you are calling them?"

"Karl will be assembling all the hardware components and loading the software applications. What I'm going to suggest next is a bit sensitive. I'm pretty sure you guys have cracked the weak encryption used in GSM cell phones. We should expect to obtain many cell calls if we can first obtain those numbers from satellite phones. Our system currently does not have decryption software for GSM but Karl can insert a code call to such a routine should you want to include it. I'll alert him you might be calling. The only other thing that would be helpful is packaging and transportation. The McChord base is near Seattle and if you could assist that would be great."

"Yesterday you mentioned something about the satellite orbits?"

"Currently we've only been tracking Globalstar based upon the data you provided to Karl. There are other providers such as Iridium. I would like those orbits. Another sensitive topic – you know I cracked the Globalstar encryption. I don't expect much of a problem with the others."

FBI James turned to AG Gates and asked, "Do we have any legal issues here? No offense Jon."

"Under the Patriot Act, we have none whatsoever. Good question, please continue Randy," was the reply.

"My last one has to do with sharing the results from your system."

Jon glanced around the table, "That's up to you. We set up a special VPN for our current system. You might consider something like that."

CIA Jim went next, "My only concern is we have a number of assets on the ground over there, mostly locals. We can't afford to compromise any."

Jon responded, "I'm very familiar with unintended consequences and share your concern. I think you need to discuss this internally and carefully such that procedures can be put in place."

Jim mouthed 'Thank You.'

Gates looked over at the National Security Advisor and asked, "What are you grinning about, Clint?"

Prescott went to serious face, "Prior to these meetings I could not have imagined this level of cooperation."

"I agree," added Shelia, "but there has been no discussion of a contract or compensation; that's almost unheard of."

Jon answered, "Not expected; not required."

Gates asked, "Any other questions? If not I have the list of action items."

The General said, "We have a number of personal and logistical topics to address. I request we schedule a meeting at the Pentagon as soon as possible. What is your schedule like Dr. Reynolds and Ms. Ruiz?"

Katie said, "First of all please call us Jon and Katie. Next I would like to know what I should bring, makeup, hair dryer, sunscreen, change of clothes?"

Everyone laughed.

"We are available. When finished here we intend to go to Seattle and test the systems Karl will assemble. One or more of you are welcome to accompany us. When they are ready, so are we."

Gates said, "Katie and Jon, we have a couple of details to address, please stay behind. Meeting adjourned and thanks to all."

After the room emptied, Jon asked, "What's up?"

"A small matter of your security clearances; stop by my office to pick up a packet of forms and questionnaires. There's no issue with the security clearance itself but we need the forms completed for our files."

The full impact of what they had done hit. There was an officer's type club on the base where they were staying. They asked to be dropped there and immediately ordered two double gold margaritas, some chips or peanuts and sunk into soft leather chairs.

"Cheers love, to us and what God holds for the future," Katie raised her glass.

They clinked glasses and Jon said, "I sure as hell hope we know what we're doing, but it feels right."

"I agree."

The second round of margaritas appeared when Katie's phone rang. She answered without a Caller ID because of the 202 Area Code. It was General Irvin.

"Mr. Gates provided your number. I hope I'm not disturbing?"

"Not at all."

"We have a series of meetings or briefings, as they're typically called, set up for tomorrow. Would you be available and could we pick you up at 8 AM? It will take most of the day. A lite breakfast will be available as well as lunch."

"We'll be ready."

An ugly green four door sedan pulled up to the house. Jon glanced at his watch and thought, 'how amazingly prompt'. They were driven across the Potomac Bridge to the Pentagon. They had seen pictures but none captured the facility adequately. The maze they endured was somewhat intimidating – probably intended thought Jon. They were escorted by two young soldiers in fatigues, wearing side arms and mirror polished boots. Another soldier was standing by a door that he opened.

Seated inside were the General, Major, Lieutenant and one other they did not recognize.

They all stood and General Irvin said, "Welcome Jon and Katie, meet Don and Charlie. I'm Sid. And this is Ralph, also known as Captain Russell. Now that that's handled relax for a few minutes; we have a visitor coming."

The door opened and a handsome man with a full head of silver hair entered. The four stars on his collars didn't need counting.

"Oh my God," Katie's hand went to her mouth, "I interviewed you almost ten years ago at the Yuma Proving Grounds."

The Chairman of the Joint Chief's smiled, "You remember Ms. Ruiz. I certainly do and you are as beautiful as ever. Congratulations on your stories of last year. And you must be the mysterious Dr. Reynolds."

He approached Jon; they shook hands.

"I wanted to stop by and offer my personal welcome and sincere appreciation. I wish all citizens of this county could emulate the two of you."

With that he turned on his heels and left.

"Quite a presence wouldn't you say?" Sid commented.

"To say the least," Jon answered.

They started with coffee and sweet rolls and then spent the day with one person after another, providing briefings:

descriptions of the environment along with pictures, current areas of conflict, the location of various bases and other facilities.

Just after lunch Sid said, "Katie, we appreciated your humor yesterday but please understand these are harsh conditions. The clothing we provide suits the terrain and conditions; the food is marginable at best but filling; there is no alcohol; the living quarters are tents but properly reinforced and windproof plus they have electricity.

"You alluded to another two people involved. What is that about?"

Katie answered, "There are two other members of our so called team that we want to include. They are native Afghans with outstanding technical and language skills who would be huge assets. We are working out their availability with Mr. Gates and will let you know."

Jon added, "If possible what would be ideal would be three tents in close proximity for us, them and the equipment."

Sid's eyebrows went up but he kept his thoughts quiet.

Both Don and Charlie had done a crash course on Jon's resume. They were in awe. Neither had said much of anything until Randy and another briefer entered the room.

"Hi, Randy," Jon said.

Sid introduced a middle aged woman, civilian dressed, "Gloria here is Randy's counterpart within DIA. And yes there is duplication, actually layers upon layers. Anyway we haven't selected the base of operations. That's what we want to do now."

Gloria inserted a CD into a nearby computer and a map of the Middle East centered on Afghanistan appeared on the large wall screen. Gloria clicked a few more times and colored dashed lines began to move across the screen.

She picked up a pointer and began, "The dashed, color lines represent the orbital paths of all known low orbit communication satellites. The different colors represent different providers such as Globalstar. This is not in real time but we can do that. Dr. Reynolds, did you bring the GPS coordinates?"

Katie dug in her bag and produced them. A few moments later two red markers pinpointed those locations.

Jon asked, "Could either you or Randy place on the map the horizon curvature either side of those locations and then add another line to the altitude of the satellites?"

Randy and Gloria conferred privately for a few minutes and then hovered over the computer controlling the screen. An additional five minutes later a series of arcs appeared.

Randy announced, "To the best of our knowledge these lines represent the geographical ranges for line of sight to the satellites."

"Really well done," said Jon, "Don or Charlie, could we add the various US or NATO base locations?"

Pleased to become involved, Don went to the computer and connected to somewhere else in the Pentagon networks. Then he pointed out something to Gloria while Randy stood by. The minutes went by; they continued typing, and then yellow circles appeared on the screen. They turned to face Jon, Katie and Sid, all were beaming.

Clearly visible were two bases or facilities located inside the satellite ranges and included the locations of Jon's GPS coordinates.

"I'm absolutely amazed," said the General, "let me check out which facility would be best suited."

Jon stood and congratulated each with a handshake and a big smile.

"I think we're ready to rock and roll," Katie stated.

Chapter 17

Jon and Katie packed their few personal effects while the driver waited. Katie gave him an envelope with their security clearance forms completed and asked if he would deliver such to Preston Gates.

The Lear took off and headed west. Six hours later they landed in Seattle with a refueling stop in Minneapolis. Despite Jon's desire to catch a cab, Karl and Marie greeted them upon landing. Stephanie gave Jon a strong hug. After arriving at the Oberling home, they all said good night and crashed.

Jon accompanied Karl to the ACT offices the next morning. The CEO greeted Jon warmly and asked when he might return and that he was missed. Jon had been on a leave of absence for almost a year. Jon replied it might be in the spring. After greeting his former colleagues, Jon joined Karl in his office to hear about the status of the two systems. Karl informed they were good to go and located in a rented storage unit south of downtown. He went on to tell Jon that Randy had called and the GSM, cell network, decryption algorithm had been provided.

Karl continued, "Jon, I've been giving a lot of thought to our system since your comments in Cabo about its limitations. An idea occurred to me after the GSM decryption code was sent by Randy and I want to pursue my idea.

"I'm sure you know there are two primary cell networks, GSM and CDMA. The latter is mainly used in this country along with GSM, but GSM is worldwide and that's what Randy provided. They are using the Kasumi A5 encryption algorithm, pretty damn weak, but that's not the issue. To pick up cell calls our receiver, as configured, won't work. It needs a different type of antenna and transponder that effectively clones the operation of a cell tower. I have no idea what cell towers will be nearby your location over there, but I thought you should know about this issue.

"Do you remember Bud Levinson at Hughes?

"Yes, he was the contract manager and lead engineer I worked with on that contract."

"Well, I ran into Bud a while ago at a conference and he's now living here in Seattle. Seems Boeing bought Hughes and they're now producing communications satellites and the associated components. I want to have a circumspect conversation with him. That OK with you?"

"Go for it."

A team of soldiers from McChord arrived in the afternoon and with military precision and efficiency packaged all the hardware components and departed with the boxes.

That evening they were seated around a fire in the Oberling living room. Marie asked, "When do you leave?"

Katie answered, "Middle of next week we think. The exact timing has not been set. We're flying back to Tucson tomorrow to pack a few things. I emailed Ricardo to set up a meeting with him and Lucinda at the Valenzuela estate the day after tomorrow. We hope that goes well."

Ricardo had provided his email address. They did not have a cell phone. Both he and Lucinda read the email from Katie and looked at each other.

"Do you think it's possible?" Lucinda asked out loud.

"I really do," Ricardo responded.

Two whirlwind days later, they were gathered in the great room at the Valenzuela estate. Lucinda and Ricardo were already there and seated when Jon and Katie arrived. Sr. Valenzuela did not appear happy. His greeting was formal and he sat waiting for someone else to begin.

Jon opened with, "Sr. Valenzuela, I made you a promise. I would not break that promise without your concurrence. Katie and I are here today to request we be relieved of that promise. I'm not going through a long preamble because I know you want the bottom line, as we say."

Valenzuela nodded, Jon continued.

"After we separated in Ciudad Obregón, Lucinda and Ricardo provided much needed assistance. Katie and I will be spending the next month in Afghanistan on a relatively secure NATO base using our technology to ferret out terrorists. We discovered their existence through the process previously described to you. With their knowledge of the area and in particular their language skills, we want them to accompany us. Finally and should you agree, Katie used her contacts within the US government that would allow Lucinda and Ricardo to immigrate and move freely across the border to maintain relationships with you and your family."

Sr. Valenzuela showed no reaction.

Speaking in Spanish Lucinda took Sr. Valenzuela's hand and spoke for many minutes. It was obviously very heartfelt and emotional. When she was done Ricardo added something, also in Spanish. While Katie understood much of what was said she showed no indication. Jon didn't understand anything.

Sr. Valenzuela said nothing. He was lost in thought. They waited.

At length he began to speak, "I live in constant fear for my family. That is our lot in these times. We have been fortunate for decades but that has come at the price of constant vigilance and occasional violence to those who seek what they don't deserve. I desire to provide opportunity for all my family, and that includes these children. I want them to become independent and make their own decisions. I fear they will become victims of the violent and corrupt world around us. When is it time to let go and trust they are ready to make their own way? I can't say with any certainty. Lucinda tells me she is ready.

"Jonathan and Katie, I trust you will guide these children and protect them as best you can. You have my blessing."

Ricardo yelled out in joy. Lucinda gave Sr. Valenzuela such a hug he gasped for air. Both Jon and Katie said, "Muchas gracias nuestro amigo."

Lucinda and Ricardo returned to Ciudad Obregón to settle school matters and pack their personal items, there weren't a lot, but not before they completed all the INS forms. Jon informed he would email the date and time for the Lear to pick them up with all the necessary clearance and immigration papers on board. He explained they wouldn't have a passport but everything else would be in place.

Jon and Katie returned to Tucson and closed up her condo packing a few personal items. Katie called her previous editor and explained she would be away for about a month and she would resume publishing her column upon their return.

"Love you Edgar," is how she ended that call.

It was departure day and as Jon later described it: a complete cluster fuck.

Someone in Mexican customs at Ciudad Obregón recognized the Lear from a previous landing and as there were no arriving passengers, only a couple of young people departing, thought it was suspicious. A team arrived for a full blown inspection. Six hours later they were allowed to take off.

The C17 Globemaster carrying the two satellite monitoring systems had been scheduled to arrive at Andrews at 2 PM EST and await passengers before continuing on to Afghanistan. Instead it flew to Lackland AFB in Texas. Jon and Katie waited at the Tucson executive terminal for the Lear, and with no information, they went to El Matador for dinner and a pitcher of Sangria. Don and Charlie waited at Andrews, fully packed, until they decided to go home until someone told them otherwise.

It finally came together the next day. After introducing Lucinda and Ricardo, they all stood on the tarmac and had a good laugh. The C17 ground crew was finished and they boarded. A crew member handed Katie a personal, handwritten message of good wishes from Preston Gates. She passed it around.

Chapter 18

Hafiz Abdul Majeed was one of the founders of the Taliban. He was a favorite of Mullah Mohammed Omar, the Commander of the Faithful or Supreme Leader of the Muslims. He was born in a small village in Kandahar Province as was Omar. They met in the early eighties as part of the Mujahedeen fighting against the Soviet occupation. During that time they also met Osama bin Laden who was fighting with another of the seven Mujahedeen factions and several CIA operatives who provided weapons and other support.

Both Majeed and Omar were fierce fighters and gained a widespread reputation for the number of Soviet tanks they destroyed. Majeed did not follow Omar to Pakistan and religious teaching after the Soviet withdrawal in 1989. He was not consumed like Omar with the strict Shia lectures of Sheikh Abdullah Azzam. Instead he returned to his family and their poppy growing business. The Western term for his role would be Sales and Marketing.

After the Soviets pulled out, Afghanistan fell into chaos and when the government was overthrown in 1992 civil war broke out with the various Mujahedeen factions and war lords fighting for control. Tucked away from Kabul in the southwest corner of the country, Majeed continued to expand the poppy growing business into opium production and worked his way up the distribution chain to Europe.

When Omar returned from Pakistan with his newly formed Taliban or 'students', Majeed received a summons. He joined again with Omar and fought alongside until the Islamic Emirate of Afghanistan was formed in 1996. He then served in several capacities: chief of intelligence, commander of the Taliban in southern Afghanistan and assistant to the Taliban financial minister Abdullah Mutmain.

Omar's rise to power was fueled by his fight against corruption and stories such as when he led armed men and freed two young girls kidnapped and raped by a warlord, then hung him from a tank gun barrel. There were many other stories and the populace flocked to his movement.

As he gained momentum many Mujahedeen leaders were opportunists and joined. Most of Omar's second and third tier leaders came from those opportunists. Many were soon promoted. Majeed was clearly first tier.

Omar's short reign of leadership was a near constant set of contradictions. He first protected the Bamiyan Buddhas and then ordered their destruction. He introduced strict Sharia law then mixed it with political expediency. He stopped the rape and sodomy of women and children and then took women out of public, out of education, and out of the workplace. Every woman was required to wear a burka. He issued a ban on poppy production calling it un-Islam while tolerating the underground economy it supported. He was pious about giving fellow Muslims sanctuary and as a result, bin Laden and Al Qaeda established many training camps.

Omar and most of his organization fled to Pakistan in 2002 after the US led invasion toppled the Taliban government. Majeed stayed behind to lead the insurgency and currently held control over a large swath of Southern and Southwest Afghanistan.

It was a very long flight, over fourteen hours plus a refueling stop in Germany. In addition to Jon and Katie's group there were eleven others. Everyone tried, most unsuccessfully, to catch some sleep. The seats were not particularly comfortable. Katie caught Ricardo wide awake, staring off into space.

"What's up Ricardo?" she asked.

"Oh, hi, I thought you were asleep."

"No chance, you looked like you were deep in thought."

"I guess I was – just thinking that Lucinda and I will be having our birthday back where we were born. No longer teenagers and what a crazy twenty years it's been. The Russians owned us when we were born; the Taliban caused my mother to lose her position at the university; Al Qaeda launched their attacks; the Americans attacked back; we ended up in Mexico; our parents died and now we're going back. How crazy is all that?"

Jon and Lucinda were listening, so were a few others.

"I'm very sorry about your parents," Katie offered.

"Thanks."

"How much and how far back do you remember? Care to share? Tell me only what is comfortable."

"Actually Katie, not very much is comfortable. What would you like to know?"

"Whatever you would like to share."

"I don't remember the Russian patrols but my parents and others talked about that. The small village where we were born, I only remember from visits to relatives and their friends. We lived in an apartment in Kabul near the university. We went to school there, but really schooling was at home, at night. My parents were the real teachers, both during the day and at night with us.

"The thing I remember the most was the trip to the US and Seattle for a whole year. It was amazing and scary at the same time. When we came back, my mother did not have her job at the university any more.

"We went back to the village, Skazar. My father stayed in Kabul to teach and earn money. My mother taught us every day. One day, after the Americans came, my father arrived in Skazar driving this small truck. We packed up very fast; I remember everyone yelling and then we drove the Khost-Gardez Pass to Pakistan. It was crazy with the truck stalling and my father yelling.

"We slept in the truck in Islamabad near the American embassy for some days; I don't remember how many. My father was gone all day, every day. My mother hid us many times. One day he told us we were going to Mexico. Lucinda's Spanish was better than mine."

A few hours later they landed at Bagram Air Force Base north of Kabul. All were amazed at the size of the base. They were escorted to temporary quarters that resembled a Motel 6 and informed a short flight in the morning would take them to Kandahar, their final destination. They were also told to go to the quartermaster's building for their issues. Directions were given.

"What the hell are 'issues'?" Jon asked out loud.

"I'm sure we'll find out; let's go there now," Katie responded.

The four walked over a quarter mile to the building and introduced themselves to a dull PFC behind a long counter. He spent almost ten minutes finding their names and locating the pre-prepared boxes containing what the Army believed to be their personal needs. He rolled them out on a cart and said, "Here you go."

The eight boxes were about two feet on each side.

"How do you propose we transport these?" Katie asked.

"No idea; how far do you have to go?"

"Temporary housing," Jon answered.

The PFC appeared to either not care or was thinking.

Chapter 19

It was still January but the temperature was in the mid-eighties. Sweat broke out on Jon's brow when they deplaned at the base airfield near Kandahar. A truck in desert camo was standing by to accept the passengers and cargo. Realizing the two systems would fill the truck they radioed for another. Jon was unwilling to depart without the systems so they stood in the hot sun. Ricardo and Lucinda studied their surroundings with intense scrutiny saying nothing.

It was mid-morning and nine and a half hours later than Washington DC. They had taken off the previous day. The landscape was brown and barren but hazy mountains could be seen to the northeast. They continued to wait for the second truck.

It finally arrived and everyone and everything was loaded. The degree of curiosity was obvious on the part of the Army drivers and crew. They drove less than two miles when a tall, perhaps over ten foot fence became visible. It seemed to stretch beyond eyesight and there were small buildings like watch towers on stilts every three hundred feet or so. No vegetation was visible in any direction. Concertina wire adorned the fence top. A cursory inspection of the trucks occurred at the base entrance. They were driven to an area towards the center where three tents stood side by side. Don and Charlie would share a single tent across the dirt road.

"Guess we're finally home," chirped Katie.

They were unloaded quickly with Jon directing the equipment to the center tent. The driver of the first truck approached them and informed: mess is over there, pointing, latrines and showers there, HQ that way to the left. He saluted slightly and left.

Katie placed a hand on each of Lucinda's and Ricardo's shoulders, "I won't say welcome home. How about you take that tent on the left? When you're settled come over here. OK Jon let's get ourselves settled."

The tents were identical as were all the others on the base: about fourteen feet wide and eighteen feet long. They were constructed of a wooden frame with tightly stretched desert camo canvas covering the frame. There was a flimsy wooden door. Inside were two twin size wood frame beds. Katie sat on one and pronounced the mattress acceptable. At the far end, opposite the door was a small sink and porta potty. At the foot of each bed was a storage trunk. There were screened windows with pull down, tied off covers on each side. That was it.

"Home sweet home," Katie proclaimed, "it's time to redecorate. I'm not sleeping that far away from you, big boy."

"You're in quite a mood lover."

"Gotta make the best, you know, I'm a desert girl and that makes me horny, remember?"

Jon just laughed and began to unwind.

Lucinda and Ricardo knocked once on the door when they heard Katie say, "Get your butt over here and help me move this bed; it's heavy."

"C'mon in I'm having a few decorating problems."

Lucinda looked at the scene and giggled. Ricardo decided it was best to say nothing.

"OK, I think we're done. Sorry my interior decorator is out of town. This substitute is almost worthless. Please sit down."

Lucinda giggled again.

Jon finally said something, "Ricardo, let's go next door and start unpacking. I've had about as much estrogen as I can handle for a while."

Lucinda giggled for the third time.

No sooner had Jon and Ricardo left when a wannabe Jeep pulled up in front of their tents. A full bird Colonel got out and approached the partially open door. He introduced himself as Col. Westin Davis, base commander.

"Sorry I wasn't here when you arrived, but welcome. Where is Dr. Reynolds?"

"Next tent over getting set up," Katie replied, "our local DA is named Wes Davis; what a small world."

"I go by Westin, not Wes," the base commander said stiffly.

Lighten up, Katie thought. Lucinda was then introduced and the commander left saying he was committed to the success of their operation and contact him for any support necessary.

Col. Davis knocked on the door of the middle tent and was beckoned to enter. Introductions were made; Jon didn't comment on the name coincidence and asked if the beds could be removed to provide more room for the equipment plus several tables and chairs for the computers. The Colonel said it would be done then received a briefing on the systems being unpacked.

Because of the canvas covering the tent frame, Jon felt confident the receiver could be set up inside, but only a test would confirm that. They stacked the boxes for the second backup system in a front corner beside the door and out of the way. Then they continued opening the receiver boxes, assembling as they went. The next concern was the type and amperage of the electricity available on the base. It came from generators and both 110 and 220 volts were available but the amperage available in the tent was unknown at that point.

The sound of a vehicle and a knock on the door signaled another visitor. This time it was two soldiers unloading two folding tables and four folding chairs. The Colonel was as good as his word. They quickly removed the twin beds and were gone, having said virtually nothing.

Don and Charlie came over from their tent to inquire how it was going and could they help out. Jon suggested they unpack the computer equipment boxes and set up on the tables. He showed them where he wanted it located. All the activity made them unaware of Katie and Lucinda watching from the doorway.

"We're hungry and it's about time for dinner. That should be quite an experience," Katie said sarcastically.

"How about half an hour; we're almost finished?"

"We'll be back."

It took less time. They stood back and surveyed the tent. Jon pronounced it 'good to go'.

The six of them walked to the Mess Hall for their first experience with Army food. They had been forewarned. It didn't turn out too bad. Neither the meatloaf nor the gravy had much flavor but was hot and filled their empty stomachs. For dessert a decent lemon bar was offered. All in all, not gourmet by any stretch but acceptable.

Col. Davis approached their table and inquired if everything was alright or did they need anything else.

Jon asked, "The electrical receptacles are all 110 volts as we had been told but what is the amperage and how many circuits are there per tent? We're ready to power up after dinner."

"I'm not really sure. Let me locate one of our electricians and send him over."

"Many thanks."

They finished and began the walk back; it was becoming dusk. The electrician was waiting for them. Jon showed him into the middle tent and explained about the systems and each component's electrical requirement.

He explained there is one circuit per tent of 20 amp capacity. The feed panel was located in a small building adjacent to a nearby pole. Nothing was underground. He went on to explain the base was set up in a hurry and many corners were cut. They were in the middle of a major upgrade as three soldiers had been electrocuted in the shower at another base, but reassured them about this base.

"Do you know the total wattage your equipment requires?"

"What time is it in Seattle now?" Jon asked.

They determined it was twelve and a half hours earlier in Seattle and Karl would be sound asleep, so Jon sent an email instead of a satellite phone call.

"I'll have an accurate answer in the morning. I think it best to wait and not power up until we know."

"I'll take a look at what we might be able to add here and come back by 0900. Does that work for you?"

"Yes, thanks and good night."

Exhausted from the travel and all the organization, they fell into a dreamless sleep. They awakened to loud sounds from vehicles plus the whirring of a drill motor close by. Jon pulled on shorts and stumbled out the door. Katie was trying to burrow beneath her blanket.

The electrician apparently decided not to wait for Jon's answer. A boom truck with a bucket and spool of wire was next to a power pole. Someone was in the bucket working at the pole top. Someone else was on a ladder at the rear of the center tent screwing a reinforcement and clamp on the existing piece of wood that already held a power line from the pole.

The electrician was standing next to the truck issuing instructions. Jon went back inside and looked at his email. Karl had responded with two numbers: using 110 volts the receiver assembly will need 2500 watts and all the computer equipment another 400 to 500. He asked for a detailed email or call when Jon was able.

The electrician was standing outside the tent having seen Katie still in bed. Jon stepped out and gave him the numbers.

He explained what they were doing, "There is available space in that panel. I am putting in a dedicate 220, 30 amp circuit to your tent. The electricity on this base has its issues but is generally reliable. The national power grid here, such as it is, we don't use for obvious reasons. It is 220, 50 cycles; therefore we have plenty of transformers for use off base. I have one for you. You'll have all the power you need."

"You guys are terrific. Many thanks."

By noon, local time, the new circuit had been installed and the primary system was ready to be activated. All of the team was watching Jon and the tension could almost be felt. The computing equipment was brought up first and the external connections tested. They could get on the Internet using Jon's SAT phone as the base did not provide access in the tents. The verified cellular service the base did provide was only accessible in the main buildings. Next the various background applications were started including the new decryption program provided by NSA Randy.

The receiver was powered on and within a minute was receiving GPS signals from five satellites. Using the information provided by DIA they carefully positioned the receiver's antenna. The Ethernet cable was connected; the workstation monitor acknowledged communication. They were ready. The tension subsided only to be replaced with anticipation.

Jon pinned up a sheet of paper showing the frequency ranges of all the networks of interest, then demonstrated how to adjust the receiver. No sooner had he dialed in the frequency range for satellite phone communications than the monitor showed a signal capture for a call in progress. He turned to the computer and clicked the combined signal decryption with audio playback icon and voices came over the attached speakers. The translation software would auto-start when the current application was complete.

Ricardo exclaimed, "That was a call demanding attendance at a planning meeting not far to the north of here." He didn't need to wait for the translation because the language was Pashto.

Don and Charlie were awestruck. Don said, "I have never seen anything like that in my life. When did you folks arrive on earth?" The translation to English was in progress and both Lucinda and Ricardo knew exactly what to do: listen again to the decrypted voices, compare with the software translation and make any edits including language nuances yielding the English language contents of the call.

Jon looked at Charlie and Don saying, "Why don't you guys inform the Colonel and our group in DC about this and report our status. Who here on the base do you guys interface with? Somebody needs this intel even though it's not complete."

Both nodded and left.

Katie took Jon's hand, leaned close for a quick kiss and whispered, "Congratulations … never a doubt in my mind."

The twins had no sooner completed their task and handed the sheet of paper to Jon than Col. Davis and three others arrived. He was out of breath. Then he reported Don and Charlie were in the command center talking with DC.

Jon handed him the edited translation. As he read it his eyes widened; then he passed it on to the other three. It read:

> *'… are required. Tonight, Obeh at 8:00. Understand?'*
> *'Who will be there?'*
> *'Everyone necessary. Plans and coordination are needed, no mistakes or surprises this time'*
> *'Will Rehmani?'*
> *'Yes and you, understood?'*
> *'Yes, signing off.'*

Jon then briefed them on the system, how the capture was accomplished and the roles of both the software applications and the twins. The Colonel was as awestruck as Don and Charlie had been. He finally got around to introducing the three who had accompanied him. They were the base's intelligence officers and Don's and Charlie's liaison group.

"When briefed on your mission I was most skeptical and reluctantly agreed to host you here. No longer, that's for certain; my apologies."

"Not necessary," Katie replied.

"Do you know this Rehmani?" Jon asked, "And where is Obeh?"

One of the intelligence officers answered, "Oh yes we do. He's Taliban and a former governor of Kandahar province committed to Omar. Obeh is a smaller town and regional center north and west of here, about 100 klicks.

"Sir? We need to discuss the possibility of an operation and soon."

"Excuse us please and congratulations. I'm very glad you're here." The four officers quickly departed.

Chapter 20

Don and Charlie returned and related the several conversations with their Pentagon bosses and also with AG Gates since they had his direct line. Taken as a whole they reported total elation and a resounding 'well done' coming from the calls. Gates informed he would communicate with the entire control group and the President.

Jon described the visit by Col. Davis and troops.

The remainder of the afternoon was spent training their DIA partners in the operation of the entire system. As human monitoring was still required, Jon, Katie, Don and Charlie divided up the day into shifts with Ricardo and Lucinda performing their translation tasks as needed.

Jon decided to try a couple of experiments. He widened the receiver's frequency range including the cell phone range to see what signals would be available for capture and confirmed the anticipated problem. The wider the range of frequencies, the more networks became potentially involved in signal capture. For many their antenna only noted a high signal to noise ratio. The consequence was the inability to capture many, probably extraneous voice and data communications, including to and from their NATO base to other similar facilities. The inability to filter and identify other communications of particular interest would be near impossible; the database would become enormous and unmanageable; and their current disk space would be filled within a very short time.

Jon showed the team the large number of signals available and explained the problem, concluding with, "This is exactly the same issue NSA and Randy deal with every day … separating the wheat from the chaff and our current antenna is not adequate for many of these frequencies."

"What do you propose?" asked Katie.

"I'm thinking about it. In the meantime let's return our focus to SAT phones only." Privately he wondered what, if anything Karl had learned.

Five attack helicopters took off from the base at 7:00 PM local time. On board were thirty-five Army Rangers. It was both an impressive and scary sight to watch. Their destination was north, northwest to Obeh. They approached at an altitude of five thousand feet. They had been briefed as a result of spy satellite photos. There was a flat area of cultivated fields alongside the Hari River just to the northeast of the town. That was selected as the drop off location. Even at that altitude their approach could be heard and to the villagers it was a familiar and ominous sound.

The onboard, long range camera on the lead helicopter showed the villagers scattering towards the buildings and disappearing from sight. Apparently one individual had other ideas for he could be clearly seen, standing in the main square, with a weapon in hand looking towards the helicopters' approach. Then he disappeared into a particular building whose location was relayed to the entire Ranger group.

.At ten feet above the ground the Rangers jumped out and deployed. The helicopters rose in altitude with their guns ready for any required support. While eleven Rangers headed directly for the suspected building, the others took defensive positions around the town. It was quiet and seemingly deserted, but they knew otherwise.

The suspected stone-clad building had front steps leading up to a carved door and steps heading downwards towards another plain door at the basement level. Three Rangers cautiously approached the upper, main door and found it locked and dark behind. Following instincts the remainder crashed the lower door and received gunfire in return that they had been expecting. They threw a flash bomb inside and waited.

It took no more than two minutes before six presumed Taliban emerged firing their weapons wildly. They were quickly cut down. Then it became quiet. Another flash bomb failed to generate any response. Then automatic gunfire could be heard a short distance away. It went on, in bursts, for several minutes.

A Ranger emerged from a side street and shouted 'Clear'. Slowly the other Rangers gathered in the square; everyone was watching the surrounding buildings. Nine presumed Taliban lay dead at an opposite side, rear entrance to the flash bombed basement. The Rangers quickly cleared the entire building. They arranged the fifteen enemy bodies and photographed each. There was no ID or any personal effects to be found after a thorough search except one cell and one satellite telephone. They took those, then bagged the weapons found and returned to the helicopter pickup point. The village remained quiet when they left, returning to base.

News of the successful operation rapidly spread but in limited and somewhat speculative, exaggerated versions. While briefing their commanding officer at DIA, Don and Charlie learned the local base intelligence officers had sent a report where they took unwarranted credit for the operation but did mention the Reynolds-Ruiz intel source without including any details. The DIA had immediately slapped a top secret classification on the report but thought it likely to have been shared with NATO command. They became very pissed off and confronted Col. Davis.

Charlie began at their private audience with the Colonel, "We need a course correction, sir. We can't allow local egos and fiefdoms to subvert mutual cooperation and I sure as hell don't need to tell you who's watching over our shoulders from Washington. That self-serving little group of intel officers you have here has a dismal track record and you know it. What's worse is that puffed up Ranger Captain believes the sun rises and sets with them.

"Sorry to be so blunt, but Don and I think you need to know and make some corrections before we lose the most valuable source of intelligence we've ever had."

Col. Davis sat very quietly and still for several minutes before speaking, "I do know what you're saying. I do see it. This man's Army is not perfect and has a lot of problems in keeping with the ugly sides of human nature. Thank you for coming forward. Allow me a few days to sort this out. I will get back to you."

On the walk back to their tents, Don wondered out loud, "Do you have any hope things will change?"

"Actually no," Charlie answered, "Why would they? Any form of cooperation or compromise is seen as weakness to these bloated egos. And besides none of them, outside of the Colonel maybe, has the brain power to see it. Or maybe they're so controlled by testosterone their mental abilities have either moved back to primal responses or their dicks. I'm disgusted and probably should have gone to work for some dot com instead of Uncle Sam."

"I think we should discuss this with Jon and Katie."

"OK."

They found Katie monitoring the system. She acknowledged their arrival but said everyone else was sleeping and suggested they get together in the morning. They agreed, left, and went to bed disgruntled.

The next morning Charlie walked to the Mess Hall and returned with an armful of breakfast for everyone including an urn of coffee. He set it all up in his and Don's tent and went to extend invitations. The tent was too small for everyone so they pulled chairs from the system tent and sat outside. No one else was around and any base noise was distant.

Charlie proceeded to explain what happened and their conversation with Col. Davis. When finished he looked around the assembled group for feedback.

Jon was visibly angry, "So we have the three stooges right here on this base. Next thing you know, we'll all be on the cover of the *Washington Post* and this project will be dead. Can anything be secret or private? Even here in the middle of an Afghan desert."

Katie was next to respond, "You're preaching to the choir my friend ... been there, done that, bought the t-shirt. It's government, including the military, at its worst, although I actually think highly of Col. Davis. Wonder the size of his cajones?"

Both Jon and the twins snickered; Don and Charlie remained serious.

Katie continued, "In the last twenty-four hours we have collected some interesting additional info. Nothing earth shattering but potentially useful. Jon will tell you I'm a believer in the frontal approach. I'm going to give AG Gates a call. What time is it in DC? I can't seem to keep track of all the relevant time zones."

Jon did the arithmetic and said she might just catch him before bed. After Katie left to retrieve the SAT phone and make her call, Jon addressed the group, "Trust that she knows first, the problems, second the strengths and weaknesses of the players and last how to manipulate, for lack of a better word, all to our benefit. I've seen her in action many times and have complete faith and trust. May God have pity on the non-believers."

The shit hit the fan less than four hours later. The midnight oil must have been burning in Washington DC. A Sergeant drove up in a wannabe jeep, got out and stood stiffly while announcing their presence was requested at 0400 hours at the base headquarters conference room. Jon confirmed and the sergeant drove off.

The so-called conference room was simply a much larger tent with a row of three, eight foot tables, surrounded by uncomfortable chairs. When they arrived, Colonel Davis sat in the middle on one side. The three intel officers sat huddled at one end. There were three others in fatigues seated across from Davis.

He greeted Jon and team with, "Thank you for coming, please take a seat. The officers over there you have previously met. Across from me are Lt. Col. Alan Bradford, my Executive

Officer, Major Ralph Johnson, our NATO Liaison Officer and Captain Bill Larson, who commands our Ranger Company."

Jon, Katie, Don and Charlie shook hands and sat down. The twins had stayed back monitoring the system.

Col. Davis began, "I won't mince any words here. Each of you knows exactly why we're here. Furthermore, any future communications I receive from our Pentagon superior officers will be positive, congratulatory and appreciative of our team's mutual efforts and successes. Am I clear?"

There were a few nods of heads otherwise it was quiet waiting for the next shoe to drop.

"I want to restate a few rules or procedures that should have been clear and practiced all along. First, all external communications will be cleared through my XO. That specifically includes anyone in Washington DC. The chain of command here runs through me." Davis was glaring at his intel officers as he spoke.

Looking directly at Ranger Captain Larsen, "And what level of braggadocio is acceptable when your Company goes into Kandahar on leave?"

Changing his focus towards Major Johnson, Davis continued, "Since when do we share our intelligence with others without specific authorization or reasons?

"How god damned stupid can you be? The Taliban and Al Qaeda have ears everywhere, from listening posts and the cafes of Kandahar all the way through European Capitals to Washington DC. Should they ever learn the source and methods of the intel that led to our recent operation two things will happen: one, we lose the most valuable current intelligence asset in our possession, and two, each and every one of you will face a court martial. Am I clear?"

The ensuing silence was deafening.

Katie broke it in even, measured tones, "I am the reason for this meeting. You are the cause of this meeting. I spoke at length with Attorney General Gates and National Security Advisor Prescott who subsequently briefed the President. Name dropping? Don't believe that for a second.

"None of you is allowed a personal agenda and recalling an old saying, 'loose lips sink ships' you may find yourselves as fish food unless you follow Col. Davis' command. Both I and our team support the Colonel one hundred percent. Quoting Col. Davis, am I clear?

"To succeed here we need to become a close knit, closed mouth team that has each other's back. Play or pass gentlemen."

Having never been spoken to before in such a manner and especially by a woman, the shock was apparent. No one said a word for almost five minutes when Col. Davis concluded the meeting.

"I'm confident we have reached an understanding. Either my XO or I am readily available to further discuss your individual roles and responsibilities. The meeting is adjourned. Good night."

Returning to their tents, Katie was almost marching alone several steps ahead. Jon, with Don and Charlie off each shoulder, was following. Don whispered, "That is the most amazing, beautiful, and scary woman I've ever seen. I don't know how you deal with it but I envy you."

"You're right on the mark Don, and thanks."

Chapter 21

It was just becoming light outside. Jon, Don and Charlie walked to base headquarters and requested use of the base's communications capability with Washington DC. It was 5 AM in Afghanistan and 7:30 PM in DC. Despite the late hour AG Gates and the entire control group was gathered in the conference room at Justice with a star shaped speaker phone in the center of the conference table. Jon thanked everyone for their off hours attendance, then opened the reason for the call. Col. Davis, his XO and one of the intel officers were on the call.

"I have a data problem here and need your input, particularly Randy, but it impacts everyone. Trying to avoid being too technical, let me summarize.

"You may remember satellite phone and data communications use a particular narrow frequency band. GSM or cell phones use another. In addition there are a number of other communication networks using additional frequencies, including yourselves for particular security reasons. Our system was designed to focus on SAT phones. When we open up the frequency range we begin seeing all sorts of extraneous signals that create a data processing, filtering nightmare. I'm certain NSA and Randy can relate to this problem. Furthermore, I don't have sufficient computer power locally to process all the signals and our antenna is not adequate for cell phone intercept. We would like to propose and discuss a solution."

Randy spoke up, "Welcome to the data overload club Jon. I can hardly wait for your proposed solution."

Before Jon was able to continue, Preston Gates said, "A couple of comments please. Jon, please convey to Katie and the rest of your team our sincere thank you. That first intel and subsequent operation were cheered by everyone here. The reports I've received indicate that, how should I say, recent personnel conflicts over there have been resolved to your satisfaction. Please continue."

"Everyone is most welcome but remember we've only been here a short time. Maybe we just got lucky. I know there's a lot to do to create a smooth running system that is effective for you, so now to that subject. Randy, would you explain to the group there what a teraflop is?" He did so.

Shelia from HS exclaimed, "You mean trillions of transactions per second?"

"Yes."

"Thanks Randy, last year when I was supporting Katie's stories, Karl and I had use of a Cray computer at 45 teraflops. Without disclosing national secrets, my guess is NSA exceeds that by an order of magnitude utilizing parallel computing. What I have here is less than two percent of that computing power plus limited data storage. We need to make some decisions that define our focus and intended targets.

"What I am proposing has four parts. First, use the existing, operational system to focus on only the SAT phone frequency band. It has the highest probability of use by upper level terrorists or drug traffickers or other bad guys. The second part suggests the activation of the backup system, already here, to focus on the GSM or cellular frequency range using new components that allow it to act as a cell tower. There's a caveat to number two: Randy, Gloria with DIA, together with Don and Charlie here, would need to work together to establish filtering criteria such that the cell phone data captured is both relevant and manageable. Both Karl and I are willing to support that effort.

"Number three is a request for additional computer horsepower here. Whether you fund it or we do is immaterial right now. You are the only ones who can deliver what we need. If you are agreeable I'll email a specification. Our poor little tent will never know what hit it.

"Fourth and last, create a shared database that would also serve as a backup. I'm guessing you already have non-Internet, secure communication systems in place that are capable of transferring very large amounts of data. Our team here doesn't have a need to know, just be able to use it. Whether it's NSA, DIA or both also doesn't matter to us.

"It would consist of a master backend database, a number of views organized by suspected groups, type of communication, keywords by language, individual names and numbers, date, time and GPS locations and a few others. It would need a user interface for management and additional queries. I have sketched out the architecture for this. That's about it.

"This isn't required by tomorrow and I'm not looking for feedback right now. Please discuss among yourselves including both Don and Charlie here. I will give them my equipment specification and database design. I have already discussed this with Karl. He is completely up to speed. Many thanks again for your time. Turns out two phones were captured by the Rangers and are now being shared with us. I need to get back to work."

Returning from base headquarters after the call, they heard first, then felt a large explosion coming from near the main entrance to the base. All three hit the dirt. Within minutes the entire base was on alert. Jon got to his feet and began running towards his tent. Don and Charlie followed. Katie and the twins were huddled together in front of the middle tent.

"Jon, thank God you're alright. Do you know what that was? Are we under attack?"

"I don't know; heard gunshots but don't hear any battle going on. We felt the concussion of the explosion. It was sudden and scared me for a moment. Are you alright?"

"We're fine."

Lucinda said, "It's like when we lived in Skazar. Bombs and explosions were frequent, some very close by."

A sergeant, driving very slow, approached and then came to a stop in front of them. He was patrolling the tent residence areas.

"Is everyone alright here?" he asked.

Katie answered, "Yes, we're fine. Can you explain what happened?"

"There is no immediate threat. We believe it was a truck containing explosives driven by a suicide bomber. Two soldiers at the outer gate were killed. That is all I know. I'm sure you will be briefed when the investigation is complete. Take care."

Katie gasped and put her hand to her mouth, "I'm so sorry."

Ricardo had returned to the system tent. He shouted, "Come, quick please."

The monitor was showing the auto-start translation software was running. It meant a new signal had been captured. They had to wait until the application completed before either reading or listening to the voices. After more than a minute Jon began to read while Lucinda and Ricardo listened. The call was in Pashto. The English translation read:

> *'It is done, God is great and may the martyr be received and blessed forever.'*
> *'Thank you for the information. I will inform others. Do we know the results?'*
> *'No. The observer was afraid to get close. I will be dealing with that soon. The explosion was at the infidels' main gate. I pray many were killed and our lost brothers watched together with God.'*
> *'Keep watch and keep me informed.'*

Ricardo reported the computer translation was essentially accurate but he made a number of edits, like changing *watch* to *alert.* Then Don rushed the final translation to base headquarters. Jon gave Charlie his equipment specification and database design notes as promised.

"Katie, it's time we did some analysis of the captured phones and the meager data we've collected so far. Let's do it in our tent. Ricardo, would you trade off with Lucinda and monitor the system today?"

"OK."

Katie turned on the captured SAT phone and pressed Menu to obtain the phone's number. She read it aloud twice while Jon scanned the list of seven unique numbers from the calls they had previously obtained. It matched one of the numbers from the first call. The fact the other number from that call was listed three times, once after the operation and again that day, meant its owner was still alive. That confirmed the other number had been the sender on the first call and the receiver on today's call.

Katie suggested they send CIA Jim an email with all the numbers and try to see if there's some logic to the prefix, like Michael did for Mexico. Jon agreed and went to his personal laptop open on the storage trunk at the foot of the bed. An Ethernet cable had been run from the center system tent. As he composed and sent the email, he was reminded to check in with Michael and Corinna.

Next they obtained all the recent numbers called or received in the captured phones and added them to the list as a separate category. They found several additional matches to the numbers on the list of seven.

"Here we go again lover, another spider's web is emerging."

"You're probably right, but I need to be thinking differently here, not like in Cabo where we were picking up a fairly steady stream of calls. Because of our location we're not getting anything from Kabul, too far away. In this part of Afghanistan SAT phones are few, even with the sizable population of Kandahar. From the couple of times we dialed in the GSM frequency range it would seem most locals use cell phones if they have a phone at all. Logic tells me only the Taliban, Al Qaeda and perhaps heroin dealers would be using SAT phones and within those groups probably only the top echelon."

"I see where you're going. Another reason for those groups to be using a satellite phone is need for communication from remote places with no coverage of any kind. Let's try to assign some kind of summary label to the subject matter of the eleven calls we've intercepted. I've already deleted the calls we made on our SAT phone."

"Good idea."

Ricardo burst in without knocking. He was beyond excited, "You are not going to believe what just happened. You're not going to believe it."

They waited for him to explain.

"A call was just made between Mexico and here. The prefix of one of the numbers is 916."

"Guzman!" Jon exclaimed.

Ricardo handed him the translation. Katie looked over his shoulder as they read it.

'This is sinaloa666 calling. Let me speak to Majeed.'
'Majeed speaking, what is purpose of call.'

'Plans have changed, Karachi no longer trusted, will be eliminated. Need to deal direct and find replacement shipper. Have two possible to be discussed in person. I send trusted one to Karachi for meeting at Movenpick Hotel. Send one of yours. Twenty percent is additional to you, your answer?'

'Name the Karachi person then say what has happened.'
'Abu Bakr. I have been cheated both product and money.'

'Thank you, it is you. Meeting to be when?'
'Soon, you tell me when now or call back.'
'I will call back.'

When they looked up Lucinda was standing in the doorway, "We thought you would be more surprised."

"I am and I'm not," Jon responded, "The translation seems somewhat choppy."

"The language was Dari and your translator is not as good as for Pashto, but everything is there. It also might be the translator who was with Guzman. We listened to both the Spanish and his Dari. Also, we think someone with the name Majeed was high up with Mullah Omar when the Taliban was in power, but we don't remember anything specific."

"Now that is really interesting," Katie said and then asked, "Where the hell do we go now with this information?"

"I'm not really sure. Let's think about it. In the meantime you and I finish what we started. Maybe Ricardo and Lucinda can start unpacking the second receiver. I'm hoping to receive some additional computer equipment so we can put it in operation."

"OK."

They had no sooner gotten back to labeling the calls when Jon's email dinged. It was from CIA Jim. They both wondered what he was doing up at that hour in DC. Together they read his response.

> *'Still awake when your email arrived, couldn't get to sleep because I immediately recognized the location of one number: Islamabad, Pakistan. I call our field officer there often on his satellite phone. Same prefix 292. The others were all issued based on Kandahar. I can't be more specific. Hope this helps and you are doing well. Contact me anytime. Bye for now, Jim.'*

"You know sweetheart, this is looking more like a jigsaw puzzle instead of a spider's web."

"I was pretty good with those as a kid, at least preteen."

"We need either Don or Charlie to do some research for us. I'm going to see if they're back yet."

They had returned after delivering the translation and sat in on a meeting. They explained the truck bomb had been owned by a service company in Kandahar and made the occasional delivery to the base, therefore it was familiar. Someone, likely Taliban, had stolen the truck. The soldiers at the outer gate house were suspicious because of the truck's speed of approach. They had signaled for it to stop and when it didn't they fired at the front tires and radiator. About 50 yards from the soldiers it exploded killing them instantly. It was witnessed by a third soldier inside the gatehouse. He was injured but would survive. From the translation, the consensus at the meeting was it was retaliation for the recent operation.

"What can we do for you?"

"First I'm asking for you two to trust us because I'm going to ask for a favor without disclosing the reasons."

"I'm on board," replied Charlie. Don nodded affirmatively.

"Thank you. We need some answers that will require research on your part. I expect your research will raise some eyebrows and probably questions. Katie and I don't want you to lie for us."

"What do you need to know?" asked Don.

"The background and perhaps current whereabouts of an Afghan named Majeed who was part of Omar's Taliban government. I don't have the rest of his name. In addition, who might this same person call in Islamabad, Pakistan? That's it for now."

"I see what you mean by need to know. We will get right on it." They left the tent.

Chapter 22

It was two days later and only three calls had been intercepted. They were very short and back and forth between two new Afghan based numbers because of their prefix. The purpose was to exchange information about an ongoing military engagement in a region north and east of Kabul. Estimated casualty figures were exchanged. Both Bagram and NATO were mentioned.

The second receiver had been assembled and tested, waiting for its new components without any connection to the computing equipment. The tent had been slightly rearranged to accommodate and it was very tight inside. Charlie knocked on the door then opened it requesting Jon and Katie come over to their tent. Lucinda was asked to monitor and Jon went to his tent and rousted Katie who was napping.

They sat on Don's bed facing him and Charlie on his bed.

Don began, "Our research efforts lit up some people we don't want to involve. You can expect some blowback real soon. Here's what we know.

"His full name is Hafiz Abdul Majeed and he was one of the founders of the Taliban and reputed to be Mullah Omar's favorite and trusted subordinate. His family grows poppies on over eighty hectares southwest of here in Helmand Province about the location you first identified. The family is believed to have added heroin manufacturing to their growing operation. He is suspected to be his family's contact for distribution and is also suspected of running a smuggling operation into Europe. The file is classified at a National Security Level and not available to us. The contacts we used have all been compromised.

"From the DIA files that our clearances allowed access to, the person he most likely called is none other than Omar himself. I have a summary file here that describes Omar's current organization, called the Quetta Shura, its location, which is a fairly short distance southeast of here, and the makeup of its organization. Short version, they control the Taliban from there and our people believe the Quetta Shura is protected by the Pakistani government in Islamabad.

"Hope all this helps and you have our backs. We might need it."

Both Jon and Katie felt terrible and said so. They promised to explain as soon as a call they knew was coming had been captured. They requested patience. Both Don and Charlie sort of shrugged; they didn't appear to be very happy.

Back in their own tent Jon and Katie looked at each other. Katie came and sat on Jon's lap giving him a tender hug and said, "I think you or we made the right decision; too many unknowns, too much opportunity for speculation, and possible unilateral actions."

"We really need that Guzman call back to close the loop. We also need to call Michael. Let's see, what time is it there? Minus fourteen and a half hours, let's stay up late to make the call."

"I know a great stress reliever we can do until then."

Michael and Corinna were just finishing breakfast when the call came in. They were happy to hear from Jon and Katie as it had been several weeks. They had received some news from Karl. They played general catch-up for several minutes before getting down to business.

"Tell us how the system is working; what have you learned, and in particular, did you capture a Guzman call a few days ago translated from Dari?"

"Yes to the latter but first I have a lot of news to share. We decided to just keep accumulating and organizing until we heard from you, but my cop's instincts are running rampant. First off Guzman and his minions are really not as smart as they think. There has been a regular changing of their numbers, but as they don't control the assignment, they always get a 916 prefix. From all the various calls we've captured, they must believe they're safe from us with these changing numbers, dumb-dumbs.

"It would appear that El Chapo and El Mayo are very tight again. Seems Zambada is operating out of Panama and their partnership is in three areas. Guzman is providing product transportation into the US. I'm not certain how but have my suspicions. Zambada is receiving and trans-shipping to Guzman product through a port operation in Panama. And together they are fighting an underground war against a group known as Los Zetas in northern, central Mexico. We have it all pretty well documented. There have been no references to us of any kind. That's the best news of all.

"About that call you mentioned, yes, we got it and read the translation. Our take was it was between the same two from a month ago and the Pakistani is trying to screw Guzman. Good luck with that, even in this corrupt world."

Katie asked, "Michael, your use of 'cop's instinct' makes me wonder if you're really itching to do something with all the data you've organized. I know you so well."

"And so do I," chimed in Corinna.

"Guilty as charged," replied Michael, "but I'm willing to wait until you return and we can discuss together. How are the two of you doing with your data capture and what about the twins? I haven't spoken to Valenzuela in quite a while."

Jon answered, "Lucinda and Ricardo have been invaluable. They are safe and seem to be happy but there's something going with them about being back in their birth country.

"The satellite phone calls have been few and far between but we've obtained some very useful intel. If Karl and Marie don't know what you described to us, please include them. Would you believe what we're doing here will connect to your efforts there like pieces of a jigsaw puzzle. Stay tuned"

"I really have no doubts and yes I'll give Karl and Marie an update. Good bye and take good care."

Two things happened almost simultaneously: a sergeant drove up to deliver a sealed envelope and Ricardo shouted, 'the call is happening now'. Jon accepted the envelope and stuffed it in his back pocket then went into the system tent. Katie was already there and both Don and Charlie heard the shout and came over. They were crowded in the tent waiting for the call, capture and subsequent processing to finish.

Jon took that opportunity to open the envelope and read the typed message inside. It was a request from AG Gates for a conference call the following morning at 6:30 local time. No greeting, no salutation, more like an order.

The wait for the translation was not long but Lucinda and Ricardo were allowed the time to listen to the voices in Spanish and Dari, and then make their edits. The final translation read as follows:

'This sinaloa666, greetings'

'This is Majeed calling to confirm. I will attend meeting suggested. Using your calendar the date will be February 12 arriving night before. Check-in name is same as previous shipper.

'You will not regret this. I will send Jose to meet along with confirming introduction and written agreement and banking instructions.'

'Ciao.'

Lucinda preempted the question, "Yes, the last word was 'Ciao' as in Italian."

Jon handed Don the delivered note from Gates. He read it and passed in on then asked, "Was that the call you have been expecting?"

"Yes and it's now time to brief the two of you. Thanks for your trust and patience. You will see why. Let's all go to Mess and get together after."

As they walked to the Mess Hall, Jon and Katie held hands then lagged behind. She spoke quietly, "Do you have a plan in mind that I don't know about? How can we brief them without some kind of plan? I'm confused and all of a sudden you're not sharing."

"Really sorry sweetheart, I'm trying hard not to slip back into old habits. You have a good point but we owe Don and Charlie something before the call. My brain is so convoluted right now I can't see the end game."

They stopped, letting the others continue on. Katie looked Jon directly in the eyes, "How many times have I told you knowing the outcome is highly overrated? Of course it's unclear, we've barely started and both of us knew who might be involved. Are you surprised? I'm not. Let's give them a brief overview and gauge what we disclose either during or after the call. I do believe Gates has our backs."

"Good advice. I'm actually hungry."

Chapter 23

They assembled once again in the base headquarters' conference room with a secure phone on the center table. Don and Charlie were on one side; Jon and Katie were opposite with Col. Davis next to Jon. Washington DC was scheduled to initiate the call. They waited. No other base personnel had been included.

The phone rang. Col. Davis reached out, pressed a button and announced their collective presence.

Preston Gates' distinctive voice came through clearly but Davis adjusted the volume slightly.

"Good evening or good morning there. We have the entire control group assembled here at Justice. Thank you for joining us. Before getting to matters at hand, I want to inform that Col. Davis has been included in this group without any reservations but no one else under his command.

"Jon, first we would like to inform that your requests and suggestions from the previous conference call have been accepted, conditionally."

"What conditions, Mr. Gates?" Jon responded.

"Among our group are concerns that the level of cooperation has somehow been altered and not in a positive direction. I am acutely aware of your distrust of government from last year's events. Katie, I'm asking you, have you reverted back to your previous positions, and if so why?

"Before you answer let me say we are aware of your request for information through officers Don Taylor and Charlie Strand. We also have been informed of the communications with Jim here in this group."

In a calm and direct voice, Jon answered instead of Katie, "I will respond, at least initially, as the decision to obtain information without consulting this group first was mine. My research request to Don and Charlie was prefaced by asking for their trust because I did not want to push them into any lies. I anticipated intense curiosity and potential scrutiny of their inquiries. What they do not know they cannot disclose and as officers in the Army they are required to follow direct orders, meaning full disclosure. The needed research required sources outside this control group and we don't need or want the inclusion of anyone else lest it become published in the media. That also includes the email exchange with Jim.

"Don and Charlie agreed to my request understanding the potential peril and for that I'm grateful. At this point they have been fully briefed. I have my own conditions to extend that briefing to you."

Katie spoke up, "Before Jon lists his conditions I have a few words to add. I won't stand still and allow Jon to take the heat for excluding you. I support that decision and here's why. Mr. Gates, we trust you completely; we are choosing to trust this group you have assembled; do we trust the government in general? You know the answer is absolutely not. Please consider for a moment what might have happened if Don and Charlie had been forced, under orders, to disclose everything they now know. Not only would this project be dead and on the front page of the *New York Times*, but you would have had many, many self-serving civilian and military bureaucrats on your doorstep. To quote a well-known authority, 'we're not paranoid; we know they're out to get us'."

"Jon, what are your conditions?"

"Any sources Don and Charlie may have used face no consequences; the information we would disclose remains within this group in absolute secrecy; our team will require your resources, consultation and assistance but not your direction."

After a momentary pause Gates responded, "Have a cup of coffee; we will call back shortly."

Katie offered to get coffee from the Mess Hall. Col. Davis said there was some closer by and he would get it. Don and Charlie thanked Jon who nodded but remained silent. The hanging clock at one end of the conference tent showed thirteen minutes had elapsed when the Colonel returned juggling five cups of coffee. At the twenty-one minute point the phone buzzed. Davis answered as before.

Gates said, "Thank you for waiting, we're ready to continue and Jon, your conditions are acceptable."

"Sir, is there any chance you could come up with a world map to view and a marking pen?"

Sheila suggested they use her laptop and the in-room projector. It was operational within a few minutes projecting the world on a blank, beige colored wall.

She offered, "I can use my mouse to annotate as you might need."

"That's what we need," Jon responded, "Alright let's put some X's on the map. First Mazatlán, next Panama City on the Pacific, then somewhere in the middle of the Arizona, Mexico border. Please remember I'm picturing your map in my head. Place a large X in the middle of Europe, let's say France.

"Next let's switch to Asia. Put three X's in Pakistan, one in Islamabad, then on Karachi and the other one on Quetta. Last put X's on us near Kandahar, northwest of Kabul and the last in the middle of Helmand Province to the west of us. Let me know when you're ready."

Shelia came back with a ready after confessing she had help identifying some of the locations. A soft chuckle was heard in the background.

Jon continued, "Before we add any lines or other annotations, let me say first, this is not like an advanced mathematics exam where proof is required. I am not attempting to prove anything, simply using logic and instincts based upon almost a year of data capture and analysis up to and including present day. And the focus is drug trafficking. We can discuss the ongoing insurgency here in Afghanistan and how it interrelates in a separate conversation.

"From the X in the center of Helmand province, draw a line with an arrowhead pointing at the X in France. Next draw another such line to Karachi and then across the Pacific to Panama. From Panama keep the line going to just east of

Mazatlán and end it at the Arizona border. Let's return to Afghanistan when you're done." They waited simply looking at each other, somewhat impatiently.

Preston Gates announced they were ready.

"What you're looking at is the flow of heroin produced by the family of Hafiz Abdul Majeed. I'm certain, by now, he needs no further introduction for you."

National Security Advisor Prescott interrupted, "Do you have documentation of this?"

"Yes we do, but remember proof is not part of this conversation. That could come with relative ease. Does anyone have any questions at this point?"

Randy asked, "Can I assume the database you are requesting will contain this documentation?"

"Yes."

With no further questions, Jon continued, "Shelia, if you could change color let's draw a triangle. Let me know when."

"Go," she said.

"The triangle starts at the X in the middle of Helmand Province, goes to Islamabad, then to Quetta and back to the start."

"It's done."

"That's the communication path between Majeed and Mullah Omar. Using that same color draw a line from Helmand Province to Mazatlán with a double headed arrow. That's the communication path between Majeed and Guzman. A meeting has been set up between Majeed and a Guzman representative. We know where and when.

"There have been several additional communications we have provided to Col. Davis that relate to the Taliban insurgency but they would only detract from the point of the map you're viewing. You now know what we know; our thanks, Shelia."

"You're most welcome."

An indiscernible buzz of conversation could be heard over the phone. It went on for several minutes before AG Gates called a halt, "I'm almost sorry to say it but in typical government fashion we need a little time to digest, absorb and of course brief the President. Allow us that. Your equipment will be en route tomorrow. Randy will be coordinating all the database development tasks. Frankly, I'm out of adjectives to describe you two and your team, so please accept a sincere thanks again. Goodbye for now."

They were seated in the Oval Office on facing couches with a burning fire alongside. In attendance, besides AG Gates and NSA Prescott were the Director of the CIA and the President's Chief of Staff. The President had been fully briefed and sat there staring at the intricate ceiling.

He spoke at last, "Beyond being great advisors and supporters, I consider each of you to be friends. Sorry to appear weak and indecisive, but there are times when I really hate this job. What you have just told me comes as no surprise and I don't have any quick solutions because none exist. In fact I don't have any solutions at all. How sad is that?

"We have our military fighting two wars with no end in sight; our interests are threatened elsewhere; a large percent of our population is using drugs to tune out their problems and desperation. Greedy people everywhere are taking advantage of the weak, and nothing will ever change in Congress because any change we propose is not in their self-interest. I probably hate this job because I can't seem to affect the needed changes."

Preston Gates overcame his initial 'feel sorry' reaction and gave his friend, the President, a pep talk.

"You are what the people wanted and elected. You have another year to go and should be at least happy not to have to worry about re-election. That also means you can almost tell Congress to go screw themselves. Any challenges to your use of executive power would occur long after you're out of office. I say use it, for the betterment of our people and to bring down those who, for totally selfish reasons, choose to flaunt our laws and take advantage of our freedoms. I'd even go so far as to presume Jonathon Reynolds and Kate Ruiz voted for you. Wake up my friend."

It seemed to work.

Chapter 24

"Katie my love, tell me the difference between a vigilante and any law enforcement officer." They were curled up together in a twin bed covered by a thin blanket despite the warm nighttime temperature.

"If by your question you're worried that our little team are vigilantes, just forget it. A vigilante operates without legal authority and if AG Gates doesn't provide legal cover, I don't know who does. And he told us under the Patriot Act and with two wars going on; we have all the legal cover we could ever want."

"But we're not employed by them or anyone else for that matter. The control group supports us of course but I'm inclined to believe they are quite happy to let us do our thing as long as they are getting what they want."

Katie chuckled, "Now you're sounding depressed-cynical. What's really bothering you?"

"Not really sure, well actually I'm pissed off at myself. When we sat there in Cabo full of patriotic righteousness and decided to contact Gates, we, or more accurately, I didn't give a moment's thought to the side effects. Here we are in this crappy tent; can't leave the base; no margaritas or an end in sight. Where was the analysis I'm supposed to be so famous for, also nowhere in sight?

"More god damn drug trails with the money financing the hidden Taliban and other fanatical terrorists; our soldiers fighting and dying for who knows what good reason. Can it ever end? I sure as hell don't know. My angry side wants to put a bullet in their fucking foreheads. My practical side knows the problems are complex and almost overwhelming. I want to be on the beach in Cabo planning our wedding and nobody knows who or where we are.

"I have serious doubts our so-called control group has any clue what to do. It's almost as if they're looking to us for solutions. How scary and absurd is that thought?"

"Wow, are you frustrated. Without that margarita, lover, keep on venting and get it all out." She snuggled closer.

Jon added, "With the equipment coming tomorrow, I just want to get it set up and running smoothly, finish training Don and Charlie, then get the hell out of here."

Because of needed equipment space and associated cable runs, it was decided to have Jon and Katie switch tents with Lucinda and Ricardo and shrink their already meager living space to accommodate some of the new equipment.

Two new calls were recorded from the number they believed belonged to Majeed; the first was outgoing and a message was left. The second was an incoming return. The other number had an unknown prefix so Jon emailed CIA Jim. The translated conversation was a notification of a shipment delay and change of payment instructions to an alternate account; the word Saudi was used.

The new equipment arrived and was delivered in a canvas covered truck from the nearby airfield. Both Don and Charlie were like kids at Christmas; Jon was nonplused. Everyone pitched in and worked late. It was almost midnight when Jon pronounced the updated equipment ready. They would bring it up in the morning.

Step by step they brought the revised system online. First the two receivers were repositioned and tuned, one to the satellite phone frequency range and the other to the GSM, cell phone range. Next the new backend server, now located in Jon and Katie's tent along with Don's laptop, was booted and the Oracle database application installed followed by adding the original workstation to the new network and to Don's laptop from the adjacent tent. Lastly, using the required credentials they connected to the NSA, Randy provided a secure network. Gloria at DIA also had access. Jon ran a number of tests then sat back with a smile, "We are good to go. Let's capture some really useful intel."

Charlie was tasked with documenting the system and staying up late to test the voice and data transfer aspects of the network with Randy and Gloria in Washington DC. All went to Mess for dinner, more meatloaf likely, then came back to their tents and crashed.

Jon and Katie were becoming bored and developing a case of 'cabin fever'. They had received no feedback from AG Gates or CIA Jim for over a week. Don and Charlie were monitoring the nearly automated system with Lucinda and Ricardo providing translation edits as needed. Jon had populated the new database with all prior data. The satellite phone calls were few and far between. The cell phone calls lit up the workstation on a constant basis as development of filtering criteria had not yet been completed. Among the mass of cell phone data being captured and analyzed was one of particular interest, triggered by a possible connection to Majeed. It was a week before the Karachi meeting. That call translation had been forwarded to Col. Davis and his intel officers.

The now familiar sergeant drove up outside, then knocked on the door. Katie opened it and accepted a handwritten note from Col. Davis. She turned towards Jon, sprawled out on his bed, and announced their presence at base HQ was requested, ASAP.

With a feeble attempt at humor, he said, "Oh, if I have to."

They walked to base headquarters and were directed to the same conference room-tent. Col. Davis, his XO, the three base intelligence officers plus one more person were already seated. The new person was in a different uniform that was vaguely familiar. He was introduced as Major Max Rutherford, USMC – a Marine. He was holding a paper that at first glance was the recent call translation Jon and Katie had provided.

He spoke up, "Pleased to meet you both. I've been marginally briefed but understand your work here has been very helpful.

"I have here the translation of a recent call you provided. Is there anything else to know?"

Jon looked at Col. Davis as if to ask, 'where's the dividing line?' Before he could answer the Colonel caught the meaning.

"Max, we have a number of competing issues surrounding their intelligence gathering. That's why my briefing was so general. I apologize but you know how it is."

"I certainly do. Allow me to provide a bit of background for Mr. Reynolds and Ms. Ruiz, is it?"

Katie answered, "Yes and please tell us how we might be helpful."

"I am attached to Camp Leatherneck, close to Lashkargah, the capital or major city in Helmand Province. That province has been the location of many, if not most, of our recent battles with the Taliban even surpassing the northwest area. We have long believed Hafiz Abdul Majeed is directing the insurgency there and receiving direction because of his known and long association with Omar. Your captured call all but confirms this. Locating him is a priority for us."

To the utter shock of everyone in the tent, Jon asked, "How would you like to take that asshole out, permanently?"

Katie whispered in his ear, "We should talk to Gates."

Major Rutherford stammered, "What am I hearing?"

Jon remained silent. Col. Davis' head swiveled between those seated at the table before telling his XO and the intel officers they were excused.

With only the four remaining, Katie asked, "How long are you able to stay here?"

"How long should I?" was the Major's reply.

Katie asked, looking into blank space in the tent, "What time is it in DC, I can't seem to keep track?"

Col. Davis responded, "Just after 4 AM."

"Major, please stay the night and meet back here at eight tomorrow morning. I know what Jon's thinking and we need to check out a few things."

Walking back to their tent she said, "God I love you, welcome back, and I'm so ready to bring it home to these guys. I'll listen to what you are thinking but I have some ideas of my own."

They talked at length then made quiet love and held each other tight until it was time to call Preston Gates. They used their SAT phone to call his personal number. He answered on the third ring. Jon made a mental note to delete this call from the system next door he knew would capture it.

Jon began by asking if a military operation could be planned and executed that was based on their intel without a gathering of either the control group or some other committee.

Gates responded, "Maybe, tell me what you have in mind." They did and received a green light with assurances the President would cover their backs.

They were awakened very early by the buzzing of their SAT phone. It was the DIA General named Gus from their control group, who promptly announced he had been given responsibility for the proposed operation. It would be military command and control, all the way. They repeated their plans and ideas provided to AG Gates and explained they had a meeting with the Marine Major in less than two hours. He informed them he had sent appropriate communications to the Major and Col. Davis with full authorization. He ended with a thanks and good hunting comment.

Katie was attempting to digest some powdered scrambled eggs and suspicious bacon from the Mess as they entered the conference room-tent at 0800. Davis and the Marine Major were already seated and offered them coffee.

"Good morning," Katie expressed brightly.

They both replied with a dull 'good morning'.

Col. Davis opened the meeting with, "Quite a batch of communications we've received. We both know bits and pieces and were told the command and reporting structure but the details seem to have been left to you. As civilians that's unusual but since the two of you are so unusual nothing much surprises me anymore." He gave them a half smile.

Jon said, "Our apologies but the current environment we find ourselves in is also most unusual. Major, I have a few questions, please.

"What do you have available for defoliation, like Napalm in the Vietnam days?"

"A similar compound that is somewhat less toxic and lethal to humans. We can use it without invoking the Geneva Convention and all the other groups that have complained and accused so loudly."

"Thanks, I doubt there are many property surveys recorded here but how can we define the boundaries of the Majeed family operation?"

"No problem, already known."

"What personnel could you place in Karachi in three days' time? It would need to be a small, for lack of a better term, hit squad."

"I can't answer that now, above my pay grade."

"That's where Majeed will be. I'm not military nor have ever been, sorry. At the risk of being presumptuous, allow me to propose the elements of this operation.

"Somewhere on the Majeed property are buildings that produce, package and ship heroin manufactured from their poppies and resultant opium. They need to be identified and destroyed. Second the fields need to be sprayed with your newer Napalm product. Majeed will be meeting with a representative of Guzman, El Chapo, I'm sure you've heard of him, in Karachi at the Movenpick Hotel in three days. They need to both be eliminated."

The Major paused for several moments while he absorbed what he had heard.

"You should have been in the military."

"Just logical I believe. Please refine our ideas as you should and let us know the final plans. Also Katie and I would like a ride on this operation. It's not our first rodeo; you might check with Captain Martin at Quantico. He knows us."

The Major smiled, "I did check on you and that little shit now outranks me."

Chapter 25

Just over thirty hours later Col. Davis summoned Jon and Katie again. He explained the current status.

"The operation is approved and underway. You have been reluctantly approved as observers in a chopper from here accompanied by my XO and two others. Thought you might want to know that first.

"The likely manufacturing buildings with several trucks have been identified from satellite images. A strike is scheduled. The spraying of the fields is on hold pending visual verification of no innocent harvesters. Apparently there was a recent incident in Helmand that included civilian casualties.

"The hit team, as you called it, will be in Karachi tonight with photos of Majeed. Whether they're captured or killed remains to be seen. We don't officially sanction assassination outside war zones. What else can I tell you? Oh, by the way Katie, a package came for you in a diplomatic pouch. We didn't open it. I'll get it in a minute."

Jon was elated, "Thank you Colonel for all your help and support; I'm most pleased. Just let us know when to board that chopper. One last item please ensure the hit team looks for and retrieves any phones from Majeed or the contact. They must be sent to you here. That is essential."

They left after the Colonel retrieved Katie's package. On their walk back Katie detoured by the Mess Hall and obtained a plastic bucket filled with ice. She covered the top so Jon couldn't see its contents. He knew something was up but had learned from almost a year's experience to control his curiosity lest the wrath of Katie emerge.

Arriving at their tents it was obvious there was nothing to do. Don and Charlie had it all under control, although Jon went in and looked over Charlie's shoulder at the monitor. Observing, Jon engaged in some wishful thinking: *after the upcoming operation, maybe they could wind down and find something dull and normal, at least normal for them.*

Katie emerged from their tent and came over to say hi to Charlie who had the current monitoring shift and to find Jon. Jon's khaki shirt overhung his shorts. She stood next to him and slid her hand under his shirt and began to rub his back, then proceeded to the other side and found a nipple. Jon jerked and looked at her.

"Follow me big boy, I have a present for you, actually more than one." They disappeared back into their tent.

Placed on top of the storage trunk were two glasses filled with a lime colored liquid in ice.

"You've got to be kidding me?"

"Not me," she replied in an innocent voice.

He took a glass, started to sip, then thought otherwise and handed it to Katie. He picked up the second glass, touched it to hers and took a large swallow.

"How in the hell did you pull this one off?"

"A reporter never reveals her sources."

But having said that she opened the diplomatic pouch to reveal two bottles, both 750 ml. They were Jose Cuervo products, tequila and mixer.

Their lovemaking was only disrupted by drinking. Whether they fell asleep from inebriation or exhaustion they never figured it out and couldn't have cared less.

The so-called hit team was actually two Navy Seals, Tony and Matt, from a ship close by in the Persian Gulf. They were chosen because of physical appearance and language skills. By coincidence they checked into the Movenpick Hotel forty-five minutes after Majeed. The five star hotel was almost in the center of Karachi, near the railway station and about a quarter mile from Chinna Creek and the Intelligence Colony adjacent to the American Consulate.

The sunken café-bar area off the lobby was quiet when Majeed took a seat and ordered iced tea. He waited patiently for

Guzman's emissary to find him as he knew would happen because he had left a message for the guest registered under the name of Abu Bakr, the former intermediary.

The Seals had identified Majeed from his picture and were seated outside the sunken area along a glass wall overlooking the pool, enjoying a soft drink. They had a wide area view.

A small, compact dark skinned man, wearing dark glasses inside, approached Majeed at his table and reached out a hand containing an envelope. Majeed took and opened it, extracted a single page and read both sides.

"Please be seated Alexandro, mucho gusto I believe you say. The introduction paper is perfect and without any further negotiation, I accept the terms. Very easy, no?"

Jose, now properly identified as Alexandro, sat across from Majeed and said, "Yes. Jefe thought you would agree."

They both spoke passable English. The Seals strained to hear the conversation.

Majeed asked, "What has become of Abu Bakr? I am certain he cheated on me also."

"No longer with us as of yesterday."

"That is good. With all the recent turmoil in our world, I am pleased you survived and distribution will become normal again. We were also impacted and have altered our banking connections. Now that we have a direct accord I will provide new account information tomorrow through Switzerland, Saudi Arabia to Dubai."

Alexandro informed, "We have a new shipping service registered in Panama that visits five ports in Asia once every three months. The schedule will vary only by weeks. Tomorrow I must obtain the port permits and berths here. For that I need your assistance."

"Should not be a problem. May I order you a tea?"

"Yes, gracias, but I'd rather have tequila."

Majeed simply smiled.

They emptied their glasses and prepared to leave. Majeed signed the bill and they agreed to meet again in the morning. As they walked towards the bank of elevators the Seals followed, chatting with each other as if oblivious to their surroundings. They had previously verified there were no security cameras either in the elevator area or in the elevators themselves.

The Seals hung back while Majeed and Alexandro waited for an elevator to open. Just as the elevator door began to close, they squeezed inside offering an apology in Urdu, the local language. Majeed nodded. The door completed its closing and the elevator began to rise. The Seals fired military grade Taser barbs into each target's chest. Both immediately convulsed, tried unsuccessfully to utter sounds and slumped to the floor continuing to twitch and shake. Alexandro gave a large arched, convulsion then became still. Majeed continued to move for another fifteen seconds until the voltage subsided. He then laid there looking up in shock. His hands were quickly zip tied together.

Tony searched for Alexandro's pulse. There was none.

"A weak heart," speculated Matt, "and convenient. Now we don't have to transport two."

The elevator came to a stop. They pressed the close button and then the button for their floor. They pulled out the barbs and verified the hallway was empty. Majeed was pulled to his feet and dragged by Matt into their room. Tony followed, struggling to pull Alexandro's inert body over the carpet.

After emptying their pockets, including room keys, they bound and gagged Majeed securely leaving him on the floor. He was still recovering from the shock. Tony retrieved a blanket and together they held Alexandro upright between them with the blanket draped over his drooped head and shoulders.

Exiting the elevator they faced a waiting passenger who looked strangely at the trio. Matt shook his head from side to side and the passenger stepped by them saying nothing. They searched both of the rooms thoroughly, leaving Alexandro in his bathroom, bagging contents of potential interest.

Returning to their room they called a predetermined number and were told a car from the Intelligence Colony would be at the garage entrance in ten minutes. Majeed had fully recovered and gotten to his knees. He was trying to say something but the gag prevented anything intelligible coming out.

"Stand up asshole, we know you speak English. We're going for a short walk and ride. Close your mouth now and breathe through your nose. Try anything and I'll raise the voltage; you'll end up like your friend. Understand?" Majeed's eyes were wide with fear but he did nod.

They passed several guests on the way out but did not draw any attention. The car was waiting with two men in Western style suits standing by the open rear door. Majeed was forced into the back seat. Tony leaned in and said, "Have a safe trip asshole. Guantanamo won't be as nice as this hotel."

The Seals returned to the hotel; formal checkout was not required as the room had been prepaid; to collect their few belongings. The same car would return in twenty minutes to pick them up for their return trip.

The next morning the body of Alexandro was discovered. The Seals had assumed the cause of death would be labeled a heart attack; however hotel security examined the body and noticed the two chest penetrations of his clothing and his skin. Also since his passport was Mexican, they called ISI, the Pakistani Intelligence Service, instead of local police.

Both the Seal's and Majeed's rooms had been cleared and the keys left behind. They were presumed to have checked out and no connection was made.

ISI contacted the Mexican Consulate and provided the passport number, explaining the assumed circumstances of death and demanded further information about the victim. The Consulate verified the passport was valid but could not supply anything else. The Mexican address on file would be contacted.

Unbeknownst to anyone in the Intelligence Colony, where the Seals had left the property taken, including the two turned off phones, the numbers associated with those phones had received several calls each. They went to message.

Their now favorite messenger sergeant arrived at 4 AM. It was still dark outside but Jon and Katie were ready and waiting. The drive to the airfield took less than ten minutes. The XO and two armed soldiers stood next to a helicopter's open door. All were quickly aboard and buckled up when the rotors spun up and they were airborne. The flight southeast took no more than fifty minutes. They hovered at two thousand feet and waited. They could see in the slowly lightening dawn, poppy fields and a cluster of buildings below.

An A10 Warthog came screaming past and fired four, of a type unknown to them, projectiles. The buildings and everything else in a 200 yard radius exploded. A few minutes later smaller secondary explosions occurred. They remained in their hover position. Daylight increased and no one could be seen on the ground, including in the poppy fields. The XO informed there was a contingency plan about to occur if the fields were determined to be uninhabited at this time of day.

An enormous, compared to them, C17 flew past slightly below their altitude. It seemed to be moving very slow and its rear cargo door was open and down. Passing over the poppy fields, rectangular bags began to fall out and when they hit the ground large puffs of what looked like smoke occurred. The C17 made three passes from different directions then flew away. There was a covering of what looked like dust over a large part of the poppy fields.

After listening to some communication in his headset, Col. Davis' XO signaled to the helicopter pilot, return to base.

In typical military fashion everything requires a post-mortem, meaning endless reports that include varied opinions, perspectives and second guessing. Col. Davis confessed as much to Katie at a rare breakfast encounter in the Mess Hall. She assured him that would stop. Reports of the raid in Helmand Province had made the news. It would have been next to impossible to prevent that. No news had leaked from the Karachi part.

Katie called Preston Gates, who called the Chairman of the Joint Chiefs of Staff, who promptly settled the matter. The files on the Majeed operation were closed, secured and all the items found by the Seals were en route to the base, in care of Col. Davis.

Jon and Katie returned to boredom the following two days until the Seal's package arrived. They decided the training was over and the only loose end was a replacement for Ricardo's and Lucinda's translation services. Don reported he was working on that and a local asset might be available. Their favorite sergeant arrived with news of the package arrival. With great anticipation they walked to base headquarters.

The Colonel and his intel officers were there and a number of items were spread out on the center table. Jon's eyes went directly to two Globalstar telephones. Bingo, he thought. One had a sticky note that read, Afghan; the other Mexican.

Seeing Jon's focus Davis said, "We haven't touched those. Go for it."

"Anything else interesting?" Katie asked.

"We're just getting started."

"I left all my notes and other documentation back at the tent with Don and Charlie. Can I take these now?"

"Go for it. We're anxious to know what you find."

Back at their tent they laid out all the call logs then turned on the new phones, verified their numbers and retrieved those called or received. Jon dictated and Katie wrote. The Mexican phone number was not on their list and without access to the database with Michael in Cabo could not be matched, but both of them recognized Guzman's number from a month ago. They did see two calls to the Tucson area code. The two captured phones had exchanged three calls.

The Majeed phone contained eight matches and thirty-one total calls over the previous ninety day period, apparently the amount that could be stored. Three calls were from the same Mexican number, identified as sinaloa666, they had previously captured. There were four other numbers not previously known. They went back to the log and Katie added dates and times to her new list.

Majeed's phone suddenly rang; it had been left on. Jon dismissed his first thought to get Ricardo and let the call go unanswered. After six rings it stopped and a flashing LED appeared on the screen. That lasted for over a minute then changed to solid, indicating a new message. Now it was time to get the twins.

The first thing they did was select 'edit the greeting' to listen to what Majeed's phone would tell a caller. It was short and terse, essentially saying leave a message and call back number. Then they retrieved the new message noting there were several others.

Lucinda answered before Ricardo, "The caller was speaking Pashto and sounded either desperate or angry. He demanded an immediate call back from this and previous calls."

There were two other messages from this same caller that were essentially the same but with increasing anxiety each time. There were two additional messages, both from sinaloa666 within the past days requesting a meeting status report. Jon began to write and without looking up, said, "We're going to roll the dice."

"What do you mean," asked Katie. He handed her a sheet of paper and she read.

'This is blank calling back. I assist Majeed and he has been very sick for over a week. He asked that I call and inform. He will try to call in a few days.'

She passed it to Ricardo. Jon waited until he finished reading before continuing, "Change or edit to make it believable, choose a name, and then please make the call. We will have the GPS tracker on and I'm almost certain you will be speaking to either Omar himself or someone very close. That number was issued out of Islamabad. Can you do this my friend?"

Ricardo exhaled a deep breath and picked up the phone, "I'll try to do my best. What if I'm asked about the sickness or anything else?"

"Say it's his stomach and you don't know anything else."

"OK."

He made the call. It was answered on the second ring. Ricardo tried to deepen his voice for some reason but managed to sound serious and convincing. He was asked for details but handled that well. The unknown person seemed satisfied and hung up. While he was unknown, his location was not. The GPS coordinates were a very short distance to the west of Quetta, Pakistan.

They got back to work organizing and cross-referencing. The twins were invited to stay and brainstorm. Half an hour later Charlie banged loudly on the door then opened it.

"What a day! I think we just hit two home runs, but first look at this," handing over a call translation.

"We made that call less than an hour ago, from this phone," Jon informed.

"Oh? I thought it very strange because Majeed's name is in there."

"Tell us about your home runs."

"First, there have been no calls on the SAT phone frequencies, but the cellular calls are lighting up the switchboard as they say. The filters are working and making the data manageable. Anyway there are a lot of pissed off people who can't contact Majeed and are worried about money. They have a meeting planned and we know where. Our second homer is a planned suicide attack on Bagram. They identified the martyr. Col. Davis already has this intel."

"What about your call?"

"We know where Omar is."

"No shit?"

Katie observed, "This is moving very fast in many directions. From past experience when the left hand doesn't know what the right hand is doing, bad things happen. DC will be waking up in a few hours. We need to rally and inform the troops."

"I totally agree," Jon responded, "Charlie, can you set it up with the Colonel and whomever. Katie will call Gates."

"Will do."

At 9:30 PM local time the base conference room was almost full as was the room at Justice seven thousand miles west where the control group was assembled. Pleasantries were exchanged before getting down to matters at hand.

Jon went first, "It is necessary for everyone to have the same knowledge. I hate the 'same page' expression. Several potential operations here are in discussion or planning and need coordination. Furthermore Katie's and my tasks here are complete and the transition or hand-off must be defined. To begin let's go around this table and report the current accumulated data or intel as you like to call it. Don, Charlie why don't you guys start us off?"

Don spent the next ten minutes providing what they knew. Col. Davis went next and informed that the Marine Major at Camp Leatherneck would be conducting an operation at the meeting location provided. He went on to report that Intelligence Officers at Bagram had identified the intended suicide bomber and were searching for him and any known associates while remaining on high alert.

Jon picked it up from there, "Both Katie and I have not had the opportunity to express our great appreciation to Col. Davis and everyone here. We are beyond impressed with the conduct of the recent operations and all the brave, professional soldiers. In particular our thanks go out to the Navy Seals for the Karachi operation. In addition to taking out those two, what they recovered has provided us with amazing insight into the Taliban's communication network. We now know the location of Mullah Omar."

The ensuing silence continued until Katie broke it, "Don't everyone speak at once." Chuckles could be heard over the phone and in the tent conference room.

She continued, "So, what's the plan, people? Where do we go from here and what's needed to bring this to an end? And I mean the flow of heroin and these soldiers home safe with their families; hard questions, harder answers."

In a somewhat resigned voice, AG Gates said, "Katie, you always seem to cut to the chase. Wrap up what you believe to be necessary and come home with your country's thanks. I'm almost sorry to tell you your activities have caught the attention of many people, including NATO.

"This gig, I think that's the right word, is not my normal role as the Attorney General. I'm here because of our special relationship. In typical DC fashion, there are many people not happy with my role and they're wanting to be in on the action, so to say. All that gets in the way of your questions and the needed answers. The seven of us in this group will try to come up with some viable plans before we see you and Jon again."

Part 5 – Problems, Politics and Plans

Chapter 26

Guzman was back hiding in his underground residence in eastern Sinaloa Province but that didn't cause any problems directing operations or casting his far flung tentacles. He no longer was confident of security at the Mazatlán safe house after the call from Jonathan Reynolds. His frustration continued to mount as he had not received any information or return calls from Majeed or Alexandro. He was hesitant to call Zambada until he could confirm dates for shipment availability in Karachi. El Mayo was currently directing their combined small army attacking Los Zetas near Guadalajara. His contracted ship was scheduled to leave Panama in a matter of days.

El Chapo was pouring himself a premixed Michalada, his favorite drink, when his phone rang. He was alone and therefore answered it himself. It was one of his trusted employees from Mexico City.

"Jefe, I have information about Alexandro. I do not think it wise to discuss on the phone; it may not be secure. That caution is in the report I want to send to you. After reading you might agree. Suggest I Fax the report to Jesús and have it couriered to you."

"Do it now. I will wait."

Michael, in Cabo would pick up and translate this phone call.

The courier arrived four and half hours later. Guzman had paced the entire time. He accepted the combination of typed and handwritten pages, and then dispatched the courier without any thank you. It was expected performance of orders.

Guzman sat down and began to read. The handwritten cover page explained the information had come from a number of sources, both official and from inquiries made by his Mexico City employee. From the reports of the Karachi hotel and the ISI, it was clear Alexandro had been Tasered and died as a result. There were no personal belongings found either on him or in his room and that included his SAT phone. The official reports made no mention of Majeed. Guzman's employee had made a lucky discovery during his contact with the hotel. After confirming there was no registration for anyone with a name of Majeed, the employee thought to ask about Abu Bakr. He was told that person had been registered the same day, but ISI claimed that same person had been discovered murdered nearby before he registered.

The employee concluded with several handwritten pages. The whereabouts of Majeed was unknown and he should be presumed dead. There would be no conceivable reason for Majeed to jeopardize his supply chain into the US by killing Alexandro. Someone else killed Alexandro and the only way that someone else could have known of the meeting was by tapping phone calls.

Guzman tossed the pages across the room as he put two and two together. His fury grew as he realized there would not be any product from Karachi for Zambada's ship to receive.

Majeed's family was resilient. They had survived many decades of war and political disorder. None had been killed in the recent raid by the US Military and their European bank account balances were intact and untouched. As true pragmatists in the heroin smuggling world it was simply a matter of adjustment, regrouping and reemergence. While the destruction of their poppy fields was painful and the opium to heroin manufacturing facilities would need rebuilding, the Helmand Province supply from other fields was enormous and they alone controlled the pipeline to North America and Europe where the customers were located.

At a family meeting they accepted their kin was either dead or captured. They knew about the Karachi meeting and the previous Abu Bakr intermediary. They also knew how to contact El Chapo as he was the only major, surviving leader in the current environment of seizures and chaos currently happening. They made the decision to contact him.

Jon spent the next two days ensuring the local database was current and complete with everything they knew. He was confident both Charlie and Don would manage the system well and the local translator was now on the base and conditionally trusted. Katie had packed their few things and was sitting on her twin bed with the door open when Lucinda approached and asked to talk.

"Katie, I know we are leaving tomorrow and the first stop is Bagram where we arrived, near Kabul. Would it be possible to visit both Kabul and our birth home village, Skazar, before we leave Afghanistan?"

Katie hid her internal reactions: 'that could be fun and exciting' plus 'how safe would that be?' She calmly responded, "Let me run the idea by Jon and let you know."

Jon left the system tent and returned to his own satisfied everything necessary had been done. He greeted Katie with, "Ready to get out of here? I sure as hell am."

She explained Lucinda's request.

Jon said, "Sr. Valenzuela would never forgive us if anything happened. No, no, no, it's too high a risk."

Calmly Katie responded, "How sure of that are you? You should think about it instead of just reacting. We do owe them something and this might be their last opportunity to visit and remember. Please, Jon, just think about it."

Having been taken down a few notches, he simply said, "OK."

He left abruptly to find Col. Davis and request a contact and an audience with the appropriate person at Bagram. He returned without saying a word. Katie was already drifting off to sleep.

The goodbyes the next day were bittersweet. Jon shook hands all around, but Katie was effusive. She gave Don and Charlie big hugs and planted a kiss on Col. Davis' cheek. Their favorite sergeant drove them and their meager gear to the airfield for the flight to Bagram and a planned overnight stay.

The four of them settled into their Motel 6 style accommodations at the Bagram Base, before Jon made an appointment with the contact Col. Davis had recommended.

At the appointment two hours later, he introduced himself as Major Tom Lewis of the Intelligence Group. Jon realized quickly that Col. Davis had provided a thorough briefing, so he cut right to Lucinda's request and described the twins' background and support. Major Lewis frowned.

"Appropriately dressed and with our own people as escorts Kabul should be safe. Traveling to Skazar is another matter as there are almost constant skirmishes with the Taliban in that northwest area because it's on one of the routes into Pakistan. Col. Davis told me we should offer our support and consideration and we are very aware of your remarkable intel. How serious are you?"

"The twins are almost insistent, being here so close to their birthplace. Is there some way to balance a quick visit with safety?"

"Let me check a few things and get back. I'll arrange an escort for tomorrow to Kabul. I understand your travel plans are westbound to Dubai; the flights from here to there are every other day. You're welcome to stay and I hope the accommodations are alright."

"Just like the Ritz," Jon joked, "And thanks."

They were squeezed into the back seat of a vehicle of unknown type or origin with two soldiers up front. Katie and Lucinda were wearing ankle length covers over their regular clothes and each had a burqa. Jon and Ricardo were wearing typical workman's clothing. The clothing had been provided at Bagram. On the short ride from the base to the city, the twins were visibly excited and gave directions to the driver towards the University area.

To Jon's and Katie's utter amazement, Kabul appeared to be unaware a war was ongoing. The streets were crowded; traffic was not congested, but everywhere women with and without burqas were about evenly split. There was a mix of buildings, both new and old. Lucinda pointed out the walk-up apartment where she and Ricardo had lived and they were allowed to get out and walk near and on the campus. It could have been any university, anywhere. The entire time the two soldiers said but a few words.

Upon their return the soldiers were thanked and Lucinda gave Katie a hug that lasted several minutes.

The news they received about the requested trip to Skazar was anything but positive. There were no planned patrols from Bagram to there any time soon. A private, escorted trip was out of the question. Their only choice was high risk and on their own, however, out of consideration an innocuous vehicle was available, should they insist. The strong recommendation was they should forego their plans.

Jon reported the news to Katie and the twins. The disappointment was obvious. Jon shrugged trying to distance himself from all their reactions, when Katie spoke up, "To hell with it! I want to do this."

Lucinda and Ricardo brightened; Jon left to see Major Lewis, returning with keys to a vehicle and two weapons.

They departed at first light for the expected six hour drive, heading north out of Bagram. The vehicle was an old Toyota four-door with oxidized paint and a lot of rust. The tank was full. Ricardo was driving; Jon and Katie in the back seat.

At Lucinda's direction they turned right on Saricha Road and followed the Panjshir River. The road was oiled gravel, not exactly paved but reasonably smooth. The river had cut a path between towering mountains; the winter scenery was spectacular. They were driving below the snow line, but just. For the next few hours they passed only a small truck and two other vehicles similar to theirs plus an animal pulled cart, but no one in their direction. They maintained forty-five miles per hour and there was no sign of habitation or hint of green on either side of the road.

They entered then passed through the small village of Angomen where a few people could be seen and then the road became all gravel and bumpy. A mile further on, their previously dissipated anxiety turned to fear. A dozen armed men were blocking the road. Ricardo came to a stop; Lucinda and Katie covered their heads with the burqas; Jon pulled his Afghan hat called a Pakol down and lowered his head. A man, seemingly in charge, approached Ricardo and his rolled down window.

They conversed in Pashto for many minutes. Ricardo remained very calm as the others could tell from his voice. They were allowed to pass through to their destination of Skazar less than an hour away.

With the armed men receding from view behind them Ricardo explained, still outwardly calm, "They were part of the Northern Alliance that supports the US and NATO here. Their actual name is the United Islamic Front for the Salvation of Afghanistan and they are enemies of the Taliban. According to our father they are remnants of the war lords and just about as bad, only different.

"There is a Taliban patrol somewhere in this area they are seeking. I told them we only wanted to visit relatives in our birth village and would be returning to Kabul in a day or two. They were satisfied."

"I was about to pee my pants," Katie responded, "Great job Ricardo." Everyone else agreed.

It was early afternoon when they arrived in Skazar and a magnificent lake could be seen on the outskirts that fed the river they had followed. Lucinda directed Ricardo to a parking place off the main road. The village was larger than Jon and Katie had expected. A number of curiosity seekers were observing as they got out of the Toyota.

Lucinda said, "Ricardo, stay here with Jon and Katie. I will be back soon."

Soon turned out to be over half an hour and because of the February temperature, they got back inside the car. But she did return and wearing a brilliant smile. She beckoned them to follow her. They walked over three hundred yards to three side by side stone and mud stucco small houses where eleven people, mostly older, stood waiting. They were all smiles and introduced as cousins, aunts, and uncles. There were two children somewhere between six and ten years old who were wide eyed in wonder.

Not knowing exactly what to do, Katie reached out her hands and several women came forward and took hold. Lucinda announced in both Pashto and English they were welcome and beds would be made up, although it would be very cramped. For the next several hours Jon and Katie sat back and watched Ricardo and Lucinda converse, presumably telling everyone what had happened, where they had been and would be going. Notwithstanding their lack of understanding the language, it was a positive and warming sight to behold.

An evening meal was being prepared and many villagers began to arrive, some with pre-prepared dishes of food. It was almost as if an impromptu celebration was in progress. The conversations and chatter were almost nonstop. Neither Jon nor Katie had any idea what they ate but it was tasty and filling.

Katie tried to engage a woman who spoke fleeting English in conversation without a great amount of success, but both seemed pleased with the effort. Then it was time for sleep. They were shown to a narrow platform covered with something like straw, smiled and said 'thank you' in Pashto, about all they knew, and then curled up close together in their clothes because of the temperature and fell asleep.

Their hosts were up and preparing Roat, a traditional sweet Afghan breakfast when they were awakened to the sound of gunshots and other automatic weapons being fired. It was close by. Jon pushed Katie under a blanket and got out of bed retrieving the weapons provided by Major Lewis. Upon emerging from their back bedroom he saw Ricardo and two others peering out from a glazed window. Ricardo explained the war here was infrequent but ongoing and it was best to stay low and inside. Jon came alongside and looked out.

He could see two women on their knees with bare heads and two men standing over them with weapons. The sound of sporadic automatic weapons continued in the background. Then a gunshot very close hit and spun one of the men standing over the women. He fell and a man Jon recognized as the one who spoke with Ricardo yesterday came running and fired another shot that obviously missed. The second man turned and fired. The recognized man took a bullet to his arm or shoulder and slumped. The shooter left the women and began to run.

Without a moment's hesitation, Jon burst through the door, took aim and put three rounds in the fleeing shooter. He went immediately to the recognized man and shouted for help.

Jon stepped away as three villagers arrived to assist. Lucinda went to comfort the women who were now off their knees and in prone position. The recognized man was able to sit up and identify himself as Aziz. He accepted the provided assistance that bound his bullet wound that had passed through his bicep without bone damage. He looked up at Jon and nodded a thank you.

The remainder of the Northern Alliance forces they had met the day before and Ricardo had described arrived in small numbers. There were no further sounds of gun fire. To Jon and Katie's amazement it was almost like a new celebration was underway. Villagers appeared with food; many including the Northern Alliance soldiers, gathered in the dirt covered cobblestones and enjoyed a breakfast. Somehow Aziz knew Jon did not speak Pashto because he approached with Ricardo in tow.

His translated message was, "Your young friend here explained who you are and why you are here. It is most fortunate for me and I extend my thank you."

Jon smiled and extended his hand which was taken.

They were approaching Bagram on their return; Ricardo was still driving when Jon said, "Now listen up, all of you. Under no circumstances can anyone else, and in particular Sr. Valenzuela, know about this trip. Is that clear?"

Dead silence followed but Jon believed they all got the message. His first task upon their return was to return the weapons with the simple explanation one had been fired for practice.

The next day they took an uncomfortable flight from Bagram to Dubai for some needed rest, relaxation and sightseeing.

They were gathered together in their four room suite, twenty-three stories up, looking down at Dubai and out at blue water.

Lucinda asked, “Where do we go from here?”

Katie suddenly asked, “Jon, what time is it in DC? Why the hell can’t I keep track, how thoughtless of us?”

Jon answered and Katie excused herself, returning many minutest later.

Katie deadpanned, “I just spoke with INS and your permanent USA Visas have been approved and will be awaiting our return. We are expected in Frankfort a week from Tuesday for a flight to Andrews.”

Ricardo yelled out and pumped his first; Lucinda smiled and clapped her hands.

Chapter 27

El Chapo's SAT phone rang; he was almost afraid to answer believing it had been compromised. Looking at the calling number, it was very close to the one used by Majeed, so he did answer. The language was not understandable and he said so in both Spanish and English. The caller changed from Dari to English.

It was from Majeed's family and before the caller could continue Guzman said, "The call might not be secure. Do you have another means of communication?"

The response was, "Not at this location, only the satellite telephone. We have no cellular coverage here. A short distance away in the city there is regular telephone service. That is all."

"Mierda! What has happened to my previous contact and when is product available."

"We do not know. He has been missing for many days. We were raided by US Marines and sustained factory and field damages but we are recovering fast."

"I must secure our communications and receive your product very soon. Alternate sources I will consider. I will call your number in 48 hours." Guzman hung up.

He next called his Mexico City employee; he didn't care who might be listening or taping.

"I want two things done immediately. Are you listening?"

"Si, Jefe."

"A complete report on the Reynolds, Ruiz pair; I mean everything past and present. Use any and all sources; money is not a problem.

"I want you to find the best available talent in Mexico to provide information or consulting about phone and other types of secure communication. Pass your results through Jesús. I will be nearby. Understand?"

"Si, Jefe. I understand and will begin now."

Their planned vacation would take them from Dubai to Athens, then to Venice and Rome before arriving in Frankfort. They made it as far as Venice. Jon's phone rang and it was Michael with Corinna on speaker.

"Sorry to interrupt but we have some bad or at least very troubling news. It involves four phone calls we captured to and from Guzman. I will email the translations including date, time and numbers involved. After you read them, call us back. It is likely Guzman knows or suspects what we've been doing."

Ten minutes later Jon had received the emails. Katie was looking over his shoulder as they read.

"So much for sliding into a safe, private life," he said with disgust in his voice.

"How do think they figured it out?"

"I don't think they have figured out how unless there's a US leak and that's very unlikely. I believe they have concluded tapping their calls would be the likely explanation for all the events.

"Look at what has happened these past two and a half months. First we tracked down Guzman himself and called him. We located his Mazatlán house that I'm sure has been raided by now along with many other interdictions from intel we've provided. How else could they explain the Helmand raid and the Karachi operation? – And in particular knowing the name Abu Bakr. Connecting it to us is an easy jump; who else is a known hacker and works with the Justice Department? We need to call Michael back and then Karl."

Instead Jon asked Michael to call Karl and email him the same translations. Because of the time zones they set up a convenient conference call for everyone.

In a calm and reassuring manner, both Jon and Katie attempted to allay their fears. They stated they were not that surprised with the news and that everyone should be calm and remain focused but ever vigilant. Guzman, the Afghans and all the other criminals around the world require secure and near instant communication capability from any location. That's why they all use SAT phones. It doesn't matter where they are and everyone knows NSA monitors cellular networks and probably so does every other Intelligence Agency from France to China.

Jon added, "We have caused a major disruption and for me it feels good. We still have our two systems in place listening. If the assholes can't use their phones, maybe they will crank up the Pony Express. That's a bad joke, but I'm going to give some serious thought to where they might go from here to facilitate their needed communications."

Michel asked, "What is your reaction to the request for files on you and Katie? That troubles me the most."

"Not sure right now. I can't imagine there's anything new and not known from last year's media dump."

Karl interrupted, "Marie just whispered to me and I agree with her. As tough as you two are and as analytical as we know Jon is, we're not buying what you're trying to sell.

"I have many concerns and questions running through my head right now. It's always best to let matters sink in and think instead of react. We all need to get together. Let's figure out where and when. Actually, the where is easy."

Katie answered first, "Thanks Karl, I do agree. Look for an email soon when we finalize our travel schedule. I could sure use some sun, warm water and great company. Love you all."

Jon added, "I second the motion."

Without any heads up to their control group in DC, Jon booked four First Class tickets from Venice to Dulles in Washington DC with a short plane change in Rome. They decided to use the flight time to discuss what would occur upon landing.

Jon shut down his SAT phone, his cell phone, his laptop and packed them away. Next he gathered the array of paper scattered around, reorganized it and put it in the back of his laptop case. Katie just watched.

Looking around, satisfied everything was put away, he reached out a hand to Katie, pulled her to her feet and kissed her full on the mouth.

"Let's go take a shower, get dressed, and take an evening gondola ride. Maybe we'll spot a good restaurant before coming back here for dessert."

Their flight had just left Europe behind; scattered clouds below them and blue Atlantic water were the only things visible out the window. Lucinda and Ricardo were almost giggling in seats 1A and 1B in the nose of the British Airways 747. Jon and Katie were across the aisle in 6C and D.

Jon had been quiet after their takeoff. Katie knew he was lost in thought and waited patiently sipping a mediocre margarita.

Jon broke the silence, "How do you see it? Our future I mean."

Without missing a beat, Katie replied, "Well, let's see. First we get married so all the sex becomes legit; then we buy a plow and horse, homestead 40 acres in Nebraska, have seven kids and live happily ever after."

Jon choked on his beer.

"Drinking beer in First Class? How tacky is that?"

"God, I love you woman."

"When do you want to get serious, now or over the Azores?"

"If truth be told probably never but that's a lot of wishful thinking. Besides you told me you can't have children."

"That's totally beside the point. You know I was attempting to lighten your mood and we have many, many decisions to make and time's a wasting."

"Then you make 'em. Just put a leash on me and I'll heel obediently."

"Yeah, right. Not now lover, but soon."

The plane was on approach into Dulles Airport. They had been awakened and consumed a small breakfast. Jon asked. "Do we contact Gates to let him know our plans?"

"What plans are those?"

"I thought we agreed you'd make them."

Katie jabbed her elbow into Jon's side and made him fake a wince, "OK I will. We're going to stay at the Watergate. I always wanted to do that."

Upon landing and obtaining cell coverage, they called the Watergate Hotel and reserved two rooms. The crowded cab ride through the morning traffic was painful. They arrived well before the check-in time, but to avoid a scene the desk clerk announced their rooms were available if they would pay for an additional half day.

Katie wondered where the famous break-in had occurred and the bellman informed her it was in the office section next door. Katie joked they should have checked in there as she had her lock picks in her purse. Jon laughed; the bellman was not amused as he had probably heard similar quips a thousand times before.

Katie called AG Gates' private number and announced their arrival and location. He barely hid his surprise then asked if the control group should meet. Katie told him yes and at his convenience, adding they had not yet decided their next stop.

The control group gathered in the same conference room at Justice at nine the next morning. Jon and Katie arrived using a taxi, begging off any courtesy transportation. The initial rounds of conversation were congratulatory and ensured everyone was up to date with the latest intel.

AG Gates and National Security Advisor Prescott informed the group an operation targeting Mullah Omar had been planned but was on hold because of very strained relations with Pakistan. Whether or not it would happen was unknown and their resignation towards international politics and affairs was obvious.

Jon spoke up when the room became quiet after all the catch-up conversations. "We asked for this meeting for three reasons. First, have there been any conclusions or decisions reached regarding Katie's last question? Second, where do we, as a group, go from here? Lastly, there are some potentially troubling new developments to report. I would like to try to answer the questions first."

As usual, the people in the room looked to Gates and Prescott for direction. It was silent for several moments.

Preston Gates was truly the backbone of the current Administration, albeit having to deal with a cesspool of competing interests. He looked directly at Jon and said, "Let me try to answer your second question, first. And in this answer please believe it has the backing of the President.

"In all my years as a prosecutor, then on the bench, and now as Attorney General, I have never before experienced a level of cooperation and mutual support as we have with this group. I will not stand still and have it dissipate. At the risk of sounding like a politician on the stump, the number and magnitude of the problems we face can appear almost overwhelming but with this group, we can make a difference, however small. Me, Clint here, the Joint Chiefs and the President want to continue the collaboration.

"To your first and Katie's question, there is no easy all-encompassing answer. Speaking to you personally and not on behalf of this or any Administration, we have our values and our Constitution that must be preserved, we pick our fights when they are threatened, currently there are many threats, and it's only through the commitment of the few that we have survived now for almost 250 years. That's only a spec in time when compared to other political entities from the past. I won't sit here and try to support or rationalize many of the decisions this government makes or has made. But we're all citizens of this amazing country and need to be aware of the threats, express our concerns along with our support, but not sit back and let others provide for our security and well-being."

"Quite a speech Preston, but I do agree with you and the relationships within this group must be maintained whether or not we have a current cause of action." Clinton Prescott stated.

Gates asked if Jon could elaborate on his third point.

Instead Katie spoke up, "Jon will provide the details but first let me give you an overview. It was bound to happen but we have information that both Guzman and the Afghans suspect we have tapped their calls, for lack of a better term. As a result the previous threat directed towards us and our team is back. We will be dealing with that issue shortly and are not looking for any level of protection from you.

"You can bet that the flow of heroin from Afghanistan along with meth, cocaine, etcetera from points south will continue to flow into this country and any interruptions we caused will soon be overcome. I'm sorry to say, my hopes are rapidly turning towards despair.

"To the point, communication or coordination or whatever between the groups that threaten us or this country and its citizens is required for their effectiveness and maybe even their survival. These people take advantage of available technologies to run their operations, just like we showed last year with money, banking and financial markets. Hit teams need instructions from those who issue the orders and provide details about the targets.

"The military and others in the attack or operations areas need similar communication capability, but most important of all they need accurate intelligence information. That's where we come in. Although I don't like it, I'm prepared to accept this so called good versus bad war has been going on for millennia and will almost certainly continue throughout my lifetime. My previous questions were probably naïve idealism.

"Both Jon and I, in another fit of naivety, sought to slide into some safe, secure, private life where none of you or the media could find us. We have awakened from that pipe dream. Our team, who you know, or at least you know about, will be meeting soon to discuss and plot the immediate future.

"Echoing AG Gates words, I for one will be advocating our proactive support of this group by providing quality intelligence when and where we can, and most importantly, disrupt their communications that threaten us and, by the way, try to have a life in the meantime."

Jon reached over and squeezed her arm. Katie did not flinch away and added, "Sorry, I sometimes get carried away, but I'm really pissed off."

"Wrong girl to piss off," Jon said to the group. He went on to explain all the recent intercepted calls and concluded with, "We know there's some shipping company that Zambada uses that will be bringing Afghan heroin to Panama and then north to Guzman and into this country. We also know the cash from sales of this will find its way back and end up financing Omar and the Taliban. It has to be stopped."

With no solutions known or even offered, the control group was quiet. Gates closed the meeting or better described as a briefing, with, "Talk among yourselves and let us know your ideas and conclusions. We stand ready to help and will keep our 'Big Ears' tuned to the networks you have identified."

As they departed, Shelia from HS approached Katie and offered a warm handshake. Jon accepted the thanks from several others.

Gates turned to Prescott and said quietly, "It's now or never. Set up a meeting with the Boss."

Leaving, Jon and Katie detoured on their way back to the Watergate by INS to pick up Lucinda's and Ricardo's documents and express their sincere thanks. Jon informed he had chartered the Lear to take them to Cabo with a brief Tucson stop over.

Chapter 28

Guzman was sitting back in the main room where Jesús lived listening on a landline to a professor and a graduate student from Instituto Tecnológico Autónomo de México (ITAM) drone on about communication systems, most of which he already knew. It was necessary to be polite and play a role to protect their cover as businessmen setting up import, export communications in Asia. Guzman had prepared a list of questions for Jesús to ask. At the end of the lecture the questions began.

It was explained that both cell phone and satellite phone communications were encrypted although only partially for cell calls; landline calls were not, and the methods of encryption varied by vendor. Cell phones could identify the approximate location of an individual's phone but didn't provide GPS coordinates; SAT phones did and for both the calling and sending devices if the available tracker was turned on. This was because of emergency location requirements. It was also explained that for a third party to listen to any cell or SAT phone call, the communication must be decrypted. This was also true of data transmitted over these phones such as email or text messages.

The questions and responses continued. Email using the Internet was only as secure as the server and provider offering the service; in essence not very. Landlines were the most private service unless a wiretap occurred but were becoming more and more limited in availability each year as people switched to mobile devices and certainly not available in remote areas.

Jesús asked Guzman's last written question which was, beyond face to face, what is the most secure and private method of communication in particular long distance.

The answer was immediate: voice over IP using your own server and a minimum 256k encryption. Jesús asked for a more detailed explanation.

"IP stands for Internet Protocol and each device connected to the Internet is identified by a unique IP address. Blocks of these addresses are assigned by region and are either static for a given user or handed out somewhat randomly by the provider of your Internet access service. Also there are providers that use Virtual Private Network (VPN) services. They will hide the location of the IP address. There's a VoIP service currently available that was developed in Eastern Europe. It is called Skype. But realize that to use this type of calling, one must be connected to the Internet. That's why Satellite Telephones are so popular because they will work from any geographic location."

The graduate student suspected there was more to this call and information request. He spoke up, "There are other old fashioned methods to secure a phone call. Also note that any ground based listening to calls would need to be in the range of an overhead satellite or a cell tower."

Jesús asked for some examples.

"Scrambling and then descrambling is one but the same device type must be on both ends. Creating your own code system is another, but again both parties must know the codes. Lastly a third party relay could work using a landline. There might be some others but we would need to do some research."

The call ended and Guzman couldn't see any simple solution to his suspected problem, unless he only made calls from city locations and that was both dangerous for him and impractical, but he did absorb the range of information the student had described. He instructed his Mexico City employee to check out the graduate student for possible recruitment. Next he dispatched a messenger to locate El Mayo and set up a face to face meeting. It was time to call the Majeed family.

Halfway around the world his call was answered promptly. They spoke in English.

"We must believe our phone calls could be intercepted. I am working on that problem. How many days from now is product available?"

"Sixteen," was the answer.

"You will be contacted by someone who knows my call name as you do. Instructions will then be provided for delivery and payment. Understand?"

"Yes, when will that be?"

"Before sixteen days."

"Very well, good bye."

They were back in the Lear en route to Cabo after two days in Tucson. Katie had stopped in at the *Arizona Star* and filed two upcoming syndicated columns with Edgar Greely and exchanged pleasantries with her former colleagues. The first addressed the separation of military families, the commitments they made and anxieties they faced. In the second column she returned to her mantra by describing in very direct language, the fundamental reasons for the demand for illegal drugs.

Lucinda and Ricardo traveled south with their new documents to spend a little time with Sr. Valenzuela and family.

No sooner had they finished greeting Michael at the San Jose del Cabo airport and loading their gear into the SUV, than Katie's cell phone rang. It was Preston Gates.

"Hello Katie, I have some news you won't like. My office was contacted this morning by a reporter from the *Washington Post.* It seems you and Jon were recognized at the Watergate Hotel and followed to and from Justice.

"We confirmed a meeting of course and said it was a follow up from last years' activity. The reporter has nowhere to go but I should caution you the next time you come to DC. This is why we housed you on base, for mutual privacy."

His voice was admonishing and Katie's response guilty, "Very sorry; pretty stupid of us, won't happen again, sorry."

He ended the call quickly without the usual pleasantries.

"Preston Gates is pissed and I really don't blame him. It never rains but it pours." She explained the Gates information to the others in the SUV.

"Son-of-a-bitch," Jon exclaimed, "and if we were followed to Dulles they now have the tail number of the airplane."

He called the charter company immediately and explained under no circumstances could they provide the flight plans. He was assured they did not do that but was also told they had no control over FAA records and several companies provided online tracking information from those and airport records by tail number, even for charters. The only good news he heard was that same Lear had made another flight between their drop off in Tucson and later pickup for the current charter.

"We are screwed," he almost yelled after the call.

Corinna gave everyone a big hug before announcing there had been two new Guzman calls. One was long and the other very short. Jon was in a foul mood before and after reading the translations of those calls. He made no comment about what he had read.

Katie asked, "Are you blaming me for the DC screw up?"

"Not at all sweetheart, we made the decision together. It's that I stupidly pretended life was normal, and realizing it's not, all the buried frustrations just boiled to the surface. I'll get it back together. Give me a hug."

Corinna could sense something was wrong; Michael took her aside and explained the call from AG Gates. Karl and Marie were due to arrive late that afternoon on a commercial flight from Seattle. Katie volunteered to pick them up so Jon and Michael could review the local database and the entries since they had left over a month ago.

About the time Michael and Jon completed their review, Karl and Marie arrived. Dinner was catered from the resort next door and they sat around the pool conversing and playing catch-up until the temperature began to fall. Serious discussions were left for the morning. Karl stated he was not tired and would like to review all the new material now. Everyone else called it a night.

After coffee and scrambled eggs with chorizo for breakfast it was time to strategize and hopefully plan their futures. Back around the pool in the sunshine Karl went first, "Let me offer a summary, then you can throw eggs at it if there's any left.

"We developed a system capable of capturing, decrypting and translating both satellite and cell calls in two very limited locations. By limited I mean the only calls we can intercept occur from line of sight satellites or in the signal range of cell towers. Our success has been based on the fact we had prior information about the general locations of the calls of interest. And now probably, the callers of interest, in both locations, have concluded someone has tapped into their communications they previously considered private and secure.

"Not needing more than one guess, Guzman believes it's us, therefore we're back in his sights. Switching to our Washington DC control group, how much additional intel they receive from either here or the Afghan base will depend on how quickly the news of our capability spreads throughout the enemies' world. Also it will depend on what alternate means of communication they might use.

"And finally, with all the publicity from last year and the media's constant attempts to put celebrity labels on Jon and Katie, it's uncertain how long we can remain under the radar. That's how I see it."

"No eggs Karl, you got it right," Katie said.

Almost wistfully Marie spoke her thoughts and feelings, "It's difficult to overcome the temptation to crawl under a comforter after a hot chocolate and pretend the world around us is something in a pleasant dream instead of what it really is: wars, drugs, greed, politics, people who struggle and on and on.

"We came together for a purpose and naively believed our success would bring some relief and we could resume normal lives, whatever normal is. But no, it's become so much more complex and I'm sorry to say, I don't even feel comfortable here in this little bit of paradise."

Corinna added, "I share your thoughts Marie. Since we have been living here, while Jon and Katie were in a war zone, I've selfishly hoped we have finally begun our retirement, securely away from a cop's world. That thought gets interrupted every time a phone call is intercepted."

Jon remained silent. Karl asked for Katie's views.

"I'm a fighter, always have been and probably always will be. Maybe that's why I got into journalism. I'm almost ashamed to admit that being angry at something actually motivates me. Thanks to Jon I've become a bit schizophrenic, adding a new personality. Love will do that. I know it's actually Martian and I'm supposed to be Venusian, but I want solutions, call me crazy."

Karl ignored Jon's continued silence and asked Michael, "Your thoughts?"

"It's very simple, once a cop, always a cop."

Corinna slowly shook her head and said, "I love him anyway." That brought a number of smiles.

"OK Jon, lost your tongue?"

"What about you Karl? I can't shake the feelings of guilt and responsibility for our current situation or maybe a better word is demise. Adding to that is I'm really disgusted and angry; sorry sweetheart, I'm not motivated the same way. Let me share something I said to Katie during a down moment in Afghanistan.

"We actually believed we had Guzman at bay and our safety was reasonably assured. Then after uncovering the heroin connection to Afghanistan through Karachi, we sat here full of patriotic righteousness, contacted Gates who set up the so-called control group and then our activities at the base.

"Assuming they mount an operation and that Zambada ship is intercepted, or Mullah Omar is taken out, who will have even larger targets on their backs? It's sure as hell not the control group. We need to wise up and get a grip. Sure we've had our successes but do any of you actually believe it has made a god damn bit of difference?"

"Cut it my friend. You're suffering from your usual malaise because you can't see or predict the outcome. Marie and I have discussed this at great length while you two were living in a tent and eating rations. Last night we talked late after reviewing everything collected here. As much as that warm comforter is appealing, we're not going there and the world we leave to Stephanie and her children is a major part of the reason. I do understand your viewpoint and maybe we really can't make a difference, but I'm with Katie, let's discuss solutions. Tell us how you see a path forward without knowing the exact outcome."

Jon didn't answer right away, glancing between his friends and the beach and water beyond, "You are special people. Why don't we focus on our personal situation first? If that can somehow be established then maybe we can address the other issues."

Before he could continue, Katie's email dinged. It was from AG Gates and only contained a link to something at the *Washington Post.* She felt herself cringe and said, "More problems, I think."

She accessed the link and brought up a featured article in that morning's edition, then read it. Using an expletive she seldom uttered, "Fucking reporters!"

That caused Jon to actually laugh, "Who just said that?"

She passed her laptop around so everyone could read the article. Full of authoritative certainty, a reporter told the world Jon and Katie had been seen in Washington DC consulting with the Justice Department about ongoing drug trafficking and prosecutions and were most probably sharing new intelligence information they had obtained. After those meetings the well-known pair had boarded a chartered jet and returned to Tucson, Arizona. They were currently either in Vail, Colorado or Baja California. Sources in the Justice Department confirmed the meetings. Updates could be expected.

No one had any comments after reading; they appeared lost in their own thoughts. Jon broke the uncomfortable silence.

"I don't view this place as a hideout; it's our vacation home whenever we want and the security is very good but can be improved. Don't forget about our consulting fee from the Columbians and their money launderers currently sitting offshore that we all share. I'm satisfied the drug proceeds seizure laws cover us."

Michael spoke up, "That article and your security comment about this place gave me an idea. Why don't I approach the resort property owners and their security company, if it's an outside contractor rather than employees, and explain the privacy we seek because of fame, they don't actually know our names, and propose we set up a fund, say $5,000 per month, half for each checkpoint gate, to be equally divided among the guards for enhanced checking of visitors attempting to pass through to here? Money does talk as we all know. Plus I think a ring of cameras that can be monitored remotely on the VPN would be a good idea."

Jon immediately responded, "Do it." The others agreed.

Karl said, "Jon, also ignoring the article for a moment, let me take it from here a short distance. I have a number of ideas or suggestions. About the money, it's beyond time you reimburse yourself for a year's worth of funding. It's way beyond time we get invited to a wedding and what's the matter with right here? It's also way beyond time Michael and Corinna took their retirement vacation, maybe a cruise or two … go see the sights you've always wanted. The rest of us are too young to retire, but we'll get there. We can set up instant communications wherever they might be when needed and this property can be their home base.

"I'm done with ACT; routine has set in and I'm tired of it. In fact their market share keeps going down because many companies are releasing new products and want to control their encryption and decryption methods in-house. The executives only care about quarterly profits and sales was never my strong suit. It's time I moved on and I don't see you going back either. At a convenient moment we should discuss where I think the ACT stock may be going. We're all about due for another economy correction and it might be worse than 2000.

"While you were in Afghanistan I began my exit. Also Marie and I have installed a very sophisticated, multi-faceted security system at home along with planning an addition with a home office and shop space.

"I've spoken with Randy at NSA a number of times and several other members of the control group. My instincts are telling me they are almost drooling to get their hands on our technology. I need to have a lawyer carefully check out my, our, employment contracts, but consider setting up a new company where all of us are members and funded with the drug money.

"It could provide a platform and environment for any future activities we might want to pursue. Lastly we need to make a couple of lists: one defining exactly who our enemies or targets are and the other defining exactly what weapons or tools we possess or can create. By weapons I mean systems like what we have here. We're not warriors; well maybe you're an exception, but the relationships we have developed provide all the firepower anyone could want."

The brilliance of Karl's vision slowly sunk in around the group and worries about the article were set aside. From the location of Jon's gaze it was obvious he was thinking, but little twitches that almost resembled a smile were at his lip corners. Then they all tried to speak at once; the comments and questions came fast and furious. Katie was laughing when she held up her hand.

"C'mon Jon, return to us, it's cold out there," Katie smacked his upper arm, but not too hard.

Jon grunted, "I was thinking about all the limitations of our system that we know and have discussed before. I'm not really sure we have anything to sell beyond some decryption software that will probably be obsolete within a year. I really like the new company as a platform idea because it allows me to properly handle the money and compensate all of you, but beyond that I can't see what this company can do other than be a shell."

Karl was not deflated by Jon's negative response, "Now that's where I come in. I said I had some ideas. You remember that I was going to contact that engineer from Hughes who's now at Boeing. Well I did. Let's have an offline conversation as it's very technical and should excite your fertile brain."

Chapter 29

El Chapo and El Mayo met in the crossroads town of Parral inside Guzman's panel truck. Guzman chose to trust their newly founded partnership by disclosing current information, the problems about the Afghan heroin supply, and the possible telephone call intercepts. There was an implied request for assistance. Zambada saw this as an opportunity but was unsure how far to push.

The one part of the technical briefing Guzman retained, above all else, was any interception must be within line of sight. In his head plans to relocate operations were underway. He felt comfortable that should he be able to manipulate Zambada and route communications through Panama, it would be safe. He had no knowledge of the monitoring system near Kandahar, Afghanistan.

El Mayo reported that the butchers of Los Zetas were mostly dead with their remnants in retreat to the Gulf Coast. Stability would be returning soon and calmer heads would prevail. Their drug trafficking business needed it; the internal wars only reduced profits.

Having been provided with complete contact methods and information, Zambada returned to Panama to arrange heroin transportation from Afghanistan. He was trying to control his elation for this recent turn of events and vowed to extract his rightful take when the heroin arrived.

At the same time as the Guzman, Zambada meeting was taking place, Jon used the connection that Randy had set up to contact Don and Charlie at the base. He chose an appropriate time of day.

"Greetings, this is Jon, happy to make contact."

Charlie was on monitoring duty and responded with, "Likewise, we miss you guys. To what do we owe the honor of this call?"

"Did you intercept a recent call from that Mexican number that included a sixteen days comment?"

"Yes, it's right here."

"Please alert Col. Davis that's when the next heroin shipment will be ready and before that a follow-up instructional call will happen. We will be monitoring on this end and will be in touch; anything else interesting there?"

"No, it's been very quiet with the monitoring but that operation where we provided the location occurred just after you left and another dozen or so Taliban are talking to Allah. Wish the whole war zone could be that easy."

"Say hello to Don, bye for now."

Jon and Karl were closeted together in the second floor office. Michael had been invited to join them and sat back quietly while Jon was reading the last of three large documents Karl had brought up on the monitor from his home computer over the VPN.

"What do you suppose is going on?" Corinna asked.

"Don't know, but I'm very curious," Katie answered.

Marie said, "I think I do but Karl has been very quiet about it. It has something to do with a replacement or add-on to our call intercepts system."

They changed the subject and Marie gave a rundown on their planned home remodel. Corinna commented she and Michael needed to decide on a home base and had thought about an apartment or condo rental in Tucson.

Katie mentioned having seen a bank owned short sale sign on a unit next to her condo on their recent stop in Tucson. She suggested they look into it. Corinna decided to make a call to the real estate broker who had sold their house.

While Katie was writing down her address a new email arrived. It was from Editor Greeley at the *Star* informing that her latest column was republished 228 times and the reviews and comments were very good.

Corinna returned and reported she could not determine over the phone exactly how many and which units were being sold. There was a Trustee involved and on site viewing and discussion was necessary.

Jon turned his head away from the monitor and looked directly at Karl, "You're actually serious, aren't you?"

"Damn straight I am, and tell me why not? I don't imagine you've forgotten the title and content of your Doctoral Thesis?"

"What was that?" Michael asked.

"Oh, some mouthful like 'Complex Fourier Transforms for Digital Signal Processing,'" Karl answered with a hint of sarcasm.

Jon scoffed and asked, "Do you want to explain your idea to Michael or have me do it to see if I got it and can remain sane at the same time?"

"You go ahead, my always sane friend."

"Michael, it seems Karl wants to buy or lease two GEOS satellites currently orbiting 22,300 miles above us and have that satellite capture, amplify any L-Band radio frequency we may desire from anywhere in the northern hemisphere and return the captured signal to us for subsequent processing and analysis. How'd I do?"

"Correct and summarized with amazing sanity."

"How far have you taken this idea? I could see many other files in your home folder."

"As they said in the Old West: a'fer piece 'n that thar valley is full of targets and critters," Karl was in rare form. Jon could only laugh.

Michael asked, "I have one comment and three questions. I actually believe you two could pull this off. What is GEOS? What can I do to help? Who's going to explain this to those girls?"

"Go ahead Marshall Dillion. It's your program."

"Let's see, to your questions, first it stands for Geosynchronous Earth Orbiting Satellite; analyze the gathered intelligence; and last, Jon here, of course."

Jon gave in, appeared genuinely excited and suggested a name for their new organization: CIP or Cyberspace Intelligence Professionals LLC. It was adopted. With that they returned to face their female partners seated around the pool.

With everyone seated, the women were attempting to decode the various facial expressions Jon, Karl, and Michael were wearing.

"I can hardly wait," Katie quipped.

Jon proceeded to repeat his description giving full credit to Karl as the source and the perp.

"Do you think it's actually possible?" Katie asked.

"The more the concept sinks in, I'm beginning to believe it but let's not kid ourselves, there's a lot to research before anything can be planned or implemented. It has the potential to overcome all the shortcomings of our current system."

He told the group about the proposed business name and the necessary structures and accounts to be set up and then gave a financial recap. Each of them would have a one-sixth member interest in the LLC. It would be organized in Delaware because of privacy and tax issues, and be capitalized with $50 million using the Bank of America.

Additionally the company Jon established in the Turks and Caicos Islands, 45172TCI Ltd, would become a subsidiary of CIP and retain $35 million as a reserve. He showed them his invoice covering all personal expenses over the last year up to date and reported it would be paid. There remained an amount of $7,455,106.15.

"Karl, Marie, Katie and I have decided the leftover amount is given to Michael and Corinna as their retirement nest-egg. Have some fun."

"No, you can't," Corinna cried. Michael was in mild shock.

"And why not?" asked Marie.

"What about you people?"

Karl replied, "Have you forgotten Jon's and my ACT stock keeps us out of the poor house? Thank you, take it, and enjoy it. Money only buys one thing and that's freedom."

Jon added, "You deserve every penny, and Karl? that was my line you used."

After all the emotions subsided they carefully went over the two lists and concluded they were representative and as accurate as possible. It was very late but Jon's call to the charter company was answered. He requested a pickup for the next day with the first stop in Tucson and a continuation to Seattle. He insisted on a different plane and was promised it would be a Cessna Citation, not the Lear.

They said good night to each other, with Jon adding, "It's going to work out, I'm almost certain. While you're away I've some code to review for our next potential weapon."

Katie quipped, "While he's doing that, I need a tan to cover this Afghan pallor."

El Mayo, Zambada, returned to Panama City and called the Majeed family contact. He also spoke English. After his credentials were established, arrangements were made to purchase product for delivery to the port of Gwadar, west of Karachi. Delivery must be by skiff as the port was currently under construction. The price would be per El Chapo's previous communication with the missing Majeed. Payment would be made upon delivery. Zambada's shipping contractor would be recognized because of its Panama registration and the arrival date of March 22nd. Also the Captain would know El Chapo's call sign. The call ended with mutual confirmations.

The system at Cabo was unaware of this call. The system at the Afghan base received and stored it. After decryption, no translation was required. Because of Jon's call, Charlie had been expecting it. When Jon awoke his email had the call transcript attached. *Got you again asshole, he thought.*

Jon immediately replied, *'Please sit on this for a short while. I'm currently in Mexico and need a day to consider appropriate response. Be in touch soon, thanks and great work guys.'*

Time zones notwithstanding, he received a response within minutes, *'OK, good hunting.'*

Katie came out in her robe to find Jon at his computer typing. She gave him a sleepy good morning kiss and asked what was happening. He gave her a detailed description of the emails, inviting her to read them. She declined and returned with coffee for both, announcing she was now awake and clear headed.

"How should we involve Gates and the group?" he asked.

"Call him, of course. We're not going to intercept that drug deal. I'll call him if you want. Besides, now that we have a plan they need to be on board."

"Go ahead, but as far as our plan is concerned let's not disclose anything until we're farther along."

Katie fumbled around to find her phone and then placed the call. It went to voicemail. Less than a minute later he called back.

"Sorry, I was on another call."

"Are you still speaking to us?" Katie asked.

"Of course," he chuckled, "I might have been a bit harsh, but it caught me off-guard. What can I do for you? Oh, I enjoyed your latest column and I'm sure many soldier's families did too. Maybe others will understand and appreciate the sacrifices our military makes."

"Thank you for that. How would you like to orchestrate an interdiction of Afghan heroin destined for Guzman and export to the US?"

"Why don't I assemble the group either later today or in the morning for a conference call? You know I tread softly and do not like to make unilateral decisions. We can discuss the interdiction details then."

"Send either Jon or I an email as to date, time and number. We'll call in."

It turned out to be 11 AM eastern time the next day, 9 AM in Cabo. AG Gates succinctly announced the purpose for the meeting.

Jon offered a cursory greeting and cut right to the chase, "On March 22nd a vessel of Panamanian Registry will be anchored in the harbor at Gwadar, Pakistan. While there a skiff or some type of small boat will deliver an unknown quantity of heroin produced by the remaining Majeed family. The heroin will be exchanged for a payment and that vessel will offload at a Zambada facility in Panama for transshipment to Guzman and his routes into the US."

FBI James asked, "How in the name of God do you guys do this? Incredible!"

Shelia asked Gates, "Is it time for the DEA to join this group? Or can you and Clint rationalize a military operation?"

Randy asked, "Does Don and Charlie know about this?"

Jon replied, "I alerted Charlie to look for the setup call. It came in 36 hours ago and we communicated again. The intel is on hold over there until this group decides what to do."

Preston Gates turned his back to the room and conferred privately with National Security Advisor Prescott. Everyone else waited.

AG Gates looked up and addressed Jon and Katie over the phone, "You brought this intelligence to us; do you have any additional comments regarding course of action?"

Katie answered, "While the idea of dealing with the DEA, either on this matter or within this group almost makes me nauseous, we're in no position to influence you one way or the other. We simply want an interdiction to occur and those involved eliminated. It's really your call."

"Alright, we'll take it from here and thank you again. Do you have anything to report about your team's plans?"

Katie responded again, "The six of us, you know who we're talking about, just spent time together addressing our situation. That included everything from new threats against us to what you are facing trying to manage a nearly unmanageable environment. We've reached a number of conclusions and are discussing possible courses of action. It potentially involves all of you, but it's still in discussion so no details on this call. Once finalized, Jon and I can be in Washington to meet for a very candid conversation."

"Let us know when," responded Clint Prescott.

Several days later Katie was standing waist-deep in the Sea of Cortez with soft waves lapping at her tanned back looking up at the center window of their estate, behind which she knew Jon was bent over a computer doing research and design. He was still Afghan winter white and her patience was wearing thin. Just then her cell phone rang and she raced to her towel on the beach to answer it on the fifth ring. It was Preston Gates who told her an interdiction operation was underway in Gwadar without providing any details. She thanked him for the info. As she was about to go and drag Jon out into the sun, her phone rang again. It was Corinna.

"Hi Katie, Michael and I have checked out the condo units. Your side by side condo and the pair to the east are owned by the same company, Hillside Properties and it's in foreclosure with Wells Fargo."

Katie interrupted, "That's who I bought from and that's who my mortgage is with."

"Your neighbors are renters, not owners and all three are for sale. We could use one for an office maybe. What do you think about that? Michael believes an appropriate security system could be installed?"

Katie was surprised as she had received nothing from her mortgage company, but then why would she? Recovering from her surprise, "I think it's a great idea. Let me talk to Jon. In the meantime gather as much detail as you can. I'll call you back."

Within forty-eight hours they had purchased three condo units in Tucson including paying off Katie's mortgage. Katie was still learning it was fun to spend money when there were no worries. Titles to all the properties were held by the subsidiary 45172TCI Ltd.

They held a conference call. Marie reported she had hired the remodeling company and they had finalized the design for modifications and additions to their home that would add an office area, a shop-storage area and several additional guest accommodations. Michael could direct installation of security devices at his convenience.

Karl reported that yesterday had been his last at ACT and he would be completing a write-up of his recent research that day. Jon told the group he completed his part and as soon as Karl could set up a meet with Bud Levinson, the former Hughes engineer, he and Katie would be on their way to Seattle.

Corinna reported that Katie's neighbors had until the end of the month to vacate and a cleaning company had been hired for the other two condo units. Michael added that Lucinda and Ricardo had returned with more questions than he could answer. They were currently housed at Katie's, one in the guest room, one on the couch. Katie laughed and gave her OK.

Karl closed the call saying he would call the engineer immediately and also would extend their VPN to Katie's condo. Michael should expect some equipment to arrive.

Chapter 30

The cargo ship under Panamanian registration was anchored several hundred yards offshore from the ongoing construction at the Pakistani port of Gwadar. It was 160 feet in length, somewhat small for an ocean going freighter, with a blue painted hull and a cabin house at the stern. A stepped ladder could be seen from the deck near the cabin house down to the waterline. Several crew members were visible on the flat forward deck.

The same Navy Seals, Tony and Matt were sitting in an inflatable boat with a powerful outboard motor, hidden among the construction zone pilings. Four other members of their team were spread out along the shore as observers. They were in radio contact. All had arrived the day before well after dark by sea.

The radios crackled, 'Newer panel van spotted arriving and parking near top of a ramp down to dock; small powerboat there; unloading three, no four, crates onto hand trucks; will advise further.'

A few minutes later, the radio announced, 'Crates loaded into stern; two men climbing aboard; two other men returning with hand carts. Should powerboat head towards the freighter; plan part 2 will be executed.'

From their vantage point the powerboat could now be seen heading directly towards the anchored Panamanian freighter. They started their outboard and took off in pursuit, announcing such over the radio.

Less than 200 feet away the Seals came up from behind and alongside the powerboat. In coordination Tony put several bullets in each of the men onboard and Matt tied a line to the powerboat. Within two minutes the crates were in the inflatable and the powerboat was cast adrift. Turning back towards shore, they sped away. Matt waved at the shocked crew members on the freighter's deck.

They did not return to their previous hiding place inside the pilings but instead turned east along the shoreline. After traveling just under two miles at high speed, they could see a grey panel van parked on a road above the rocky, deserted beach. They turned to port and came ashore.

With the help of the other Seals the crates were transferred to the van and all six squeezed in for the ride to Karachi and the Intelligence Colony next to the American Consulate. Two of the van's previous passengers were securely bound, gagged and stacked along the driver's side. Interrogations were part of their futures.

The freighter pulled anchor and departed but not before contacting El Mayo with full, observed details provided and assured him the endorsed bearer bonds would be returned.

At the same time the Marines from Camp Leatherneck in Helmand Province were underway in a separate operation to eliminate, once and for all, the Majeed family operations and connection to the Taliban.

Michael had accepted delivery of several boxes ordered by Karl. They were stacked in Katie's dining area. Lucinda and Ricardo didn't know what to think but were told 'to hang tight' Jon and Katie would arrive soon and explain.

A delivery truck also arrived at the Palmilla Estate with boxes containing the additional security equipment Michael had ordered and the driver complaining roundly of all the security checks he had to endure. Michael's implemented idea was obviously working. Jon and Katie began to unpack the equipment after providing a generous tip.

Travel plans were finalized. The Citation would pick up Jon and Katie for a flight to Tucson and a two day stay before continuing on to Seattle.

Katie opened the door to her condo and was greeted by Lucinda with an enormous smile and strong hug. Jon and Ricardo punched knuckles. Michael and Corinna arrived a short while later. They had been staying at a local Residence Inn. Katie told them, without any option for objections, they were all going out for dinner to her family's restaurant. She gave her uncle Hector a call and a heads-up. Jon could hardly wait to sip one of their famous margaritas.

Over dinner they came to several decisions: Michael and Corinna would furnish and occupy the second condo of the pair next to Katie's. The first condo of that pair would be similarly furnished for guests such as Karl and Marie. Upon the moving out of Katie's neighbors, that condo would be cleaned and turned into an office with a double sliding door installed in the adjoining wall. It would also be the near-term residence of Lucinda and Ricardo. Michael would oversee these steps and have the necessary security systems installed. Corinna announced she and Michael had booked an Alaskan cruise for July.

Both Jon and Katie strongly suggested the twins enroll at the University of Arizona and complete their studies. They were assured they would be kept in the loop of information and would receive a stipend to cover expenses until graduation as a bonus/thank you for all their help and contributions.

With all the details having been worked out and the VPN up and running, Jon and Katie were driven to the airport for the charter flight to Seattle. Karl was at the King County Airport waiting when they arrived. He explained Marie was supervising the construction crew and warned them to forgive the mess they would encounter.

It was like old home week being back at the Oberlings, and the late March cloudy and gloomy skies outside were ignored because of the company and a blazing fireplace. Stephanie had squeezed in between Jon and Katie on the sofa facing the fire; Karl and Marie were on side chairs.

Katie said, "We have an announcement, dear friends; on the flight here we set our wedding date, May 30th, the one year anniversary of meeting. Would you honor us by being best man and matron of honor?" Placing her hand on Stephanie's knee she continued, "And as you, my dear, are way too old to be a flower girl, I would love to have you be a bridesmaid."

Stephanie jumped up and whooped; tears could be seen in Marie's eyes. Karl simply said, "About time and congratulations."

Guzman took the call from Zambada and listened to the information received. After a short pause for absorption, he began almost screaming into the phone.

"Did you arrange the contact and transfer from Panama?"

"Si, El Chapo."

"Tell me how anyone could know; the suppliers would never have been the source. You are careless with our security and I won't tolerate that, understand?"

"Take a breath and concentrate on what happened," El Mayo responded with a great effort to remain calm, "I am never careless and somehow, somewhere, somebody listened to my call to your source. There is no other explanation."

Guzman began to calm and said, "I accept this is not your fault or doing. I think I know how it happened. I will contact you soon, but cannot now explain the contact method to be used."

Guzman hung up and turned to his ever present bodyguards and yelled, "Where is that report I ordered on Reynolds? Tell Carlos and Jesús to get their asses here pronto, hear me?"

He picked up the satellite telephone on the table before him and overcame the temptation to crush it to pieces.

It was the morning of the day Jon and Karl were scheduled to meet Bud Levinson. The remodeling crew was already at work and making a lot of noise. That forced Katie to get up; Jon was still snoozing. Stephanie had left for school much earlier. Katie was engaged in a many year habit of perusing the news stories of the day, or at least the headlines, on TV and the Internet. That morning she was in mild shock at what she saw on her screen and as bold headlines for each of the major US daily newspapers front pages:

"Government Spying on Americans"
"1984 is Happening Now"
"NSA Accused of Illegal Surveillance"
"Nothing is Safe or Private"

She ignored many other outrageous headlines from untrusted sources and chose to read a few she assumed would be reasonably accurate. Then she engaged in an almost desperate search for detailed information and finally sat back somewhat relaxed after finding no mention of either themselves or any supposed connections.

She heard Jon stir and then get up to use the bathroom. After listening she thought, *Men, there's no hope they'll ever close the door or can be trained. God help us all.* She then went and poured a cup of black coffee and held it in an outstretched hand as he passed by.

She beckoned him to come to her computer and read the headlines without any comment. He did so and followed several links reading the content as it appeared. He sat back down without comment and appeared to be lost in thought. Katie refilled their coffee cups. As she was about to speak, her cell phone rang. It was AG Gates.

"Morning Katie, I assume you've seen the news. Sorry to have put the group on hold but our activities need to stay under the radar until the dust begins to settle. I'll be in touch; please be patient."

She responded with a simple OK, a goodbye, hung up and then told Jon what Gates had said. He was now wide awake and focused. Karl was an early riser and coming down the stairs from the bedroom, they saw him watching TV in an alcove area off the living room.

Karl glanced up and said, "What are your thoughts, my friend?"

"Not sure right now … pretty jumbled; why don't we call Randy? Maybe he can provide some insight. By the way, Gates called and the control group is on hold, until when we don't know."

Randy answered after three rings. Karl and Jon identified themselves as being on speakerphone and asked if he could talk and would he be willing to provide some background.

"It occurred to me this morning that I might, for one or the other of you. Give me a call back number and wait an hour or so." They provided Karl's home landline.

They ate a light breakfast and waited. Just under an hour later the phone rang.

Randy explained he didn't want to talk from his office and would share what little he knew, "A disgruntled NSA employee had leaked to the Associated Press and the New York Times several thousand pages of reports and internal memos concerning NSA surveillance activities and methods since 9/11."

Randy freely admitted that in the government's frenzied attempt to identify potential threats and persons of interest, they probably stepped over the line, but had been assured by the Directors it was all covered and authorized by the Patriot Act. He told them he was unable to offer any prediction about when and where the fallout would land. Randy's last comments caused Jon to re-evaluate:

"We're so screwed up internally you cannot believe it. The competition between the groups and their bosses gets in the way of anything productive. In particular, arguments over who gets what information or intelligence that we gather are part of a constant war. It almost seems like the bosses fight to possess and control our intel, only to use it for personal advantage, usually for political gain and influence.

"Sorry to dump on you like this but I don't know how much longer I'm willing to put up with all the crap for the good of our country that seems to be falling apart."

"Thanks Randy. What you shared with us stays with us. We've been struggling with what our role might be in the future and actually came up with a plan. Gates has put us on hold because of this. Right now maybe we'll re-think it."

"You know, guys, our little control group has been a breath of fresh air for me. One side of me wants it to continue but I know how impractical that is. You probably have no idea what it has been like for Gates to hold it together while all the vultures who know a little about us and want control circle around.

"The only reason our activities weren't spread across the media is because nothing has been leaked so far, in part because of the National Security classification Prescott assigned to us and the integrity of the people. Speaking of Prescott, he'll now be the target of outrage for the so-called NSA spying with demands for his head. I've been with NSA almost ten years and watched the self-serving bloat grow every day, especially since 9/11. The President should really clean house, but he's a lame duck and they'll all be gone in about ten months anyway. The whole thing is really a pisser."

Jon pondered Randy's comments; Karl and Katie were listening but remained silent. Randy was about to say goodbye when Jon asked, "Are you aware of any private organization in the intelligence gathering business?"

"Beyond private detectives, no, why do you ask?"

"Just curious."

Randy suspected otherwise. With sincere appreciation and thanks given, Randy said goodbye and ended the call.

"What are you two thinking?" Katie asked.

Jon responded, "Too many jumbled thoughts right now. We both need some time to analyze all the repercussions."

Karl commented, "It's just business as usual in Washington DC and the dust will settle with some new scandal on the front pages, probably involving some Senator's sex life. God help us all. But I'm not canceling today's meeting or putting anything on hold while all those jerks fight it out."

Marie's reactions to Randy's comments were also uncertain regarding their plans but did agree with Karl to proceed while they followed all the media reports and perhaps consider a private meeting with Gates in the near future.

Katie said, "I'm in agreement with Karl."

"So am I," added Jon, "I almost forgot for a moment our lives will not wait around for them."

Guzman's original idea did not pan out. He had looked through his phone's call list and found the call he had received from Jon Reynolds because of the date and time. The number Reynolds had used was very similar to his own and when he called it was informed it was not in service.

Guzman found the NSA spying stories troubling and interesting at the same time. He searched all his available news sources but could not find any details including references to satellite phones. He sat back and thought about his three current problems: secure communications, sources of heroin, and methods for revenge. A day later he received a thick envelope containing the Reynolds, Ruiz file he had ordered. With great anticipation he began to read and then reread.

Chapter 31

The Executive Terminal at Boeing Field offered business services including a conference room. Both Jon and Karl had become well known because of all the jet charters that had used the facility. The meeting was set for 2 PM there.

Karl was waiting in the reception area when Bud Levinson arrived. They shook hands and entered the conference room. Jon stood up and they renewed a nearly ten year old acquaintance. Levinson was a burly man just under six feet with thinning gray-black hair, a neatly trimmed mustache, and chin beard.

He explained that after Boeing acquired Hughes in 2000 they began developing the Boeing BSS-601 spacecraft bus here in Seattle; while manufacturing remained in California, he had moved north and never looked back or south.

"The technology is changing so fast we can barely keep track and maintaining backward compatibility has become near impossible. When Karl told me you would be joining this meeting, I was mildly shocked that the former star software architect at ACT, and now world famous PhD, would be involved in Karl's plans. Is this some type of encore?" Bud said with some mirth.

Jon decided to be somewhat candid, "If you followed our activities last year, it shouldn't be a stretch to imagine our team is not too popular in certain quarters, aka drug traffickers. Karl's project, I should say, personal project of which I'm a part is all about self-preservation. We have a need to know where they are and if they're coming in our direction."

"I can understand and accept that. Before we continue, let me say this meeting and my involvement is personal and based on curiosity and respect. Karl has offered a consulting fee and has satisfied me there are no potential conflicts with ACT or anyone else. He has also satisfied me your activities are in no way illegal. I will only disclose information that is not classified. You understand my current employment.

"Karl and I have had a number of recent conversations and document exchanges, all outside my daytime job. Knowing I don't need to explain digital signal processing to you of all people and further knowing you're fully briefed on our consultations, let me respond to Karl's questions. OK?"

Karl said, "Speaking for both Jon and myself, your willingness to support us is most appreciated, please continue."

"As I understand your intention, you would like to receive, using an appropriate communications satellite or Comsat as we call them, L-Band radio transmissions of selected frequencies from both ground based sources and low orbit satellites, amplify them, filter the signals, and return the stream to you at a specific location back on earth. Further the geographical focus is the northern hemisphere but within that vast area you are able to define small targeted regions. Have I got it right?"

Karl answered, "Yes, both Jon and I have verified it is technically possible and we have the means to receive and manage the captured signals. Go on ..."

"To meet your requirements the Comsat must possess the following attributes and functionality: first and foremost a stable geosynchronous position for both the eastern and western halves of the northern hemisphere. Next an antenna array that meets your frequency band requirement and is both adjustable and capable of producing what we call a spot beam. In addition it must have sufficient amplifiers and transponders that are programmable. And last, sufficient solar panels and Xeon fuel for power and control."

Jon spoke up, "So far you haven't scared us away and our research indicates you have defined every necessary aspect."

Bud added, "You must remember that the signals you seek will be weak as the amplitude of the wave reduces with the square of the distance, but as we can talk with spacecraft millions of miles away, it's really not that big of a problem because you're telling me the end signal processing will be done on earth and not a function in the satellite itself."

"Bottom line and my last question," Karl interjected, "What might be available up there that meets our needs?"

"You two might just be in luck, let me explain. You might be aware that Hughes, before the Boeing acquisition, owned DirecTV. Earlier I told you the technology is rapidly changing. It's actually worse than that. Several countries around the world, including us, are launching, or throwing up there is a better term, many satellites in an attempt to utilize the latest and greatest advances in audio, video, Internet, and data communications. The competition is fierce. We are constantly developing new and improved satellite busses.

"For example, DirecTV, over the past five years, has contracted with us three different times for replacement satellites that meet newer and more complex specifications. It's crazy out there and you can't just send a repairman 22 thousand miles into space to install an upgrade. And what happens to the obsolete satellites? They get put into graveyard orbits."

Karl snickered knowingly; Jon shook his head in sympathy. Bud continued.

"Turns out DirecTV currently owns four satellites scheduled for deactivation next month. We are now preparing their replacements, with a new bus, for launch next week. Each meets your functionality requirements. Of the four, one has a longitude slot about 112 degrees west running roughly through Idaho and Sonora in Mexico. The other's slot is about 63 degrees east. It recently came off a lease with a Russian company. Considering that you're not interested in ocean coverage that could be perfect, although Japan and Australia would be marginal."

Jon asked for an explanation of how to program and control them. Bud told him it was accomplished by a proprietary remote connection application that had a user interface and the underlying language was Assembly.

Karl asked if there was a method to buy or lease a decommissioned satellite. Bud told him it was done all the time and their control programs were usually included. He would be happy to email contact information that should lead to the right person or department.

Jon added more questions, "Where would a receiving ground station need to be located for these satellites."

"Anywhere with line of sight for the fixed ground receiving antenna would work well, but I would recommend your locations be chosen based upon subsequent use. Of course programming and control could be done from almost anywhere with a link to the antenna."

Jon closed the meeting with, "It's been great to see you again and now there's the matter of your consulting fee. Karl? Did the two of you agree upon something?" Karl shook his head no.

"Bud, two questions; what's your fee and would you consider continuing these consultations?"

"Well, I probably have spent, including today, about thirty hours on this. What's a reasonable rate? I think Boeing pays about $150 an hour for consultants. How about $4,500? Does that work for you?"

Jon took out a checkbook and wrote a check for $45,000 dollars and then slid it over to Bud whose reaction was near shock.

"Did you forget the decimal point?"

"No and that's the value of your contribution to us. We want to continue to consult. Is that alright?"

"Without a doubt and thank you, thank you very much."

Because Jon's name was so well known and an alias wouldn't stand up long term, Karl made the DirecTV contact he had received via email from Bud. It took several call transfers to locate the right person, and after explaining his reason for calling, Karl had to provide both an email address and a call back number. It was hurry up and wait as usual.

Jon and Karl did their best to explain to Katie and Marie the results of the meeting and the DirecTV contact they were expecting but it was obvious the information was either not being understood or not properly presented. They dropped the subject and enjoyed another warm evening fire and good companionship. *What a day; what a life,* Katie thought, *but I wouldn't trade it for anything.*

The Oberling's phone rang at 6:40 AM Pacific time. It was the DirecTV return call and Karl was just up fixing a pot of coffee. He answered; the rest of the house was still asleep.

The caller summarized the reason for the call. His accent was foreign and not discernable but his English was good.

Karl responded, "Yes, we understand you are about to decommission four communication satellites, and my company is interested in utilizing their capabilities for unrelated purposes, but only the two focused on the northern hemisphere."

"What is the name of your company?"

"My name is Karl. How is my company name relevant before we discuss availability, price, terms and a contract? I was not very pleased with all the jerking around that occurred before this contact."

"We must be careful," the accented voice replied.

"And why is that?"

There was no answer to that question. The caller broke the ensuing silence, "We have been known to release our interests in obsolete satellites before. What do you propose and intend?"

"First question, would it be sale, lease and what control capabilities would be included or other conditions that would be attached?"

"We are prepared to discuss either and full control would be part of any arrangement."

"What's the price?"

"Lease or purchase?"

"Purchase preferred."

"Five million US per satellite."

"We can offer no more than three with no strings or other conditions attached."

"I will consider and call back."

Jon was dressed in shorts and a sweatshirt, his substitute pajamas, when the phone rang again he watched Karl answer and speak in a quiet voice, probably respecting the other sleepers as it was a Saturday.

He had poured a cup of freshly brewed coffee when Karl hung up the phone and looked in his direction.

"We are now, or soon to be, the proud owners of two satellites. Unfortunately there is no available transportation to and from our new resort in the sky property."

"Are you practicing to be a standup comic?"

Karl laughed and said, "Wake up, lighten up and let's have some fun."

"You're determined to get me out of my funk."

"Damn straight, good friend. I haven't had this much fun since you drug me up that mountain at your cabin."

Jon remembered and burst out in laughter, spilling part of his coffee.

"I'll clean it up," Karl said quietly, "and let's keep this under wraps until it's done."

Marie was emerging from the master bedroom as Katie was descending the stairs. They both saw the men seated in the living room sipping their coffees. Karl and Jon glanced sheepishly at each other and made the mistake of not offering a 'good morning'.

Katie addressed them, "And what are you miscreants up to? Actually, don't answer that; I'm not awake yet."

Marie agreed and fetched coffee for herself and Katie.

Seated together in total silence, Katie could no longer stand it, "Alright boys, give it up."

Jon chose to adopt some of Karl's recent humor, "We just bought a resort in outer space, but the accommodations are not up to your standards and any remodeling or other decorating would be difficult at best, sorry."

"You're so full of crap it's beyond belief."

Karl interjected, "You know Jon, I love to have you guys visit, but certain influences tend to disrupt our serenity."

"Get used to it," Marie responded with a snicker.

The actual purchase took over two weeks with many lawyers involved. The purchaser was finally determined to be the new LLC Jon had formed, and because of various notifications required to a number of international agencies the forms were intrusive. Jon and Karl figured out quickly these agencies had little or no authority and only wanted to record the satellite ownership and operational responsibility in their records.

It was almost the middle of April before the sale was completed and the control application was exchanged for the purchase price, all done through escrow. Karl contacted Bud Levinson and inquired whether he was available for further consultation.

Part 6 – Intelligence Gathering

Chapter 32

April 2007 was a month many would just as soon forget but there were a few 'stellar' moments. Guzman was putting out feelers to heroin sources in Southeast Asia, using his Mexico City employee whose communication capability was considered secure. No stable source had been established. Those potential suppliers were not willing to change their existing distribution channels even considering the US prices El Chapo was offering. Guzman wondered how far spread was any knowledge of recent events. He had given up on the Afghans.

The shipping company Zambada had used was no longer interested in any contracts or consignments. They apparently had decided the risk versus reward equation didn't balance. While his cocaine and ephedrine shipments continued north to Guzman unabated, El Mayo could almost sense storm clouds forming.

Guzman continued to fume over his inability to locate and eliminate his adversaries. He created an almost laughable code system, where one word means something else, to use during phone calls. More than one of his intimidated and fearful subordinates wondered if El Chapo was losing it.

The media continued to be almost obsessed with NSA's purported wrong doing; in part because of the ongoing Congressional investigations but primarily because of their negative editorial viewpoint of the current administration and the President in particular. Clinton Prescott had testified although it was not made public. The President remained silent, claiming to allow investigations to continue and allowing 'the chips to fall where they may'. Both Prescott and Preston Gates remained out of the spotlight.

The military surge in Iraq and the ongoing insurgency in Afghanistan continued to claim American lives, but was lightly reported in the media. Katie decided to alter the focus and topics of her weekly column.

Lucinda and Ricardo were totally bored with their classes and longed for some relevant action in their lives. Even Michael and Corinna were becoming bored with having next to nothing to do except view the sights and lights of Tucson from their hillside perch. They had made a retirement 'bucket list' but each item was out in the future and their focus was on tomorrow.

The Oberling household was the rare exception. The remodel was nearly complete; Katie and Marie had become almost like sisters. They appeared busy and fully occupied almost every day, coming and going. The upcoming wedding was being planned in every detail. The 'boys', Karl and Jon, as they were increasingly referred to, were in their element.

They took advantage of every opportunity that Bud Levinson's schedule would allow. Bud had recommended a certain satellite dish, and the remodeling crew had reluctantly installed it on the just completed roof of the addition. Karl almost got into it with that crew as he and Jon attempted to run the needed cables from the dish to the main house. The remodeling crew boss took Marie aside to complain. She told the 'boys' in no uncertain terms to wait, as it would only be a few days and they could then set up any 'man-cave' they wanted.

With imaginary tails between their legs, Karl and Jon slammed the front door on their way out. After a long walk, several beers at an attractive looking pub, they decided to return but not with any apologies whatsoever.

While they waited they divided duties with Jon focusing on the control software and Karl on the backend database they would need to collect signals from expected geographical regions.

From the satellite purchase they had received the command and control application including source code along with a manual describing each function of the satellite bus. That bus was proprietary to Boeing, but with Bud's guidance its subsystems were clear and control functions identified.

The bus consists of the following subsystems: Command and Data Handling, Communications System and Antennas, Electrical Power System, Propulsion, Thermal Control, and Attitude Control System Guidance.

Bud had also provided a detailed schematic of the satellites themselves showing each of the components the bus controlled. The satellite had multiple antennas so more than one spot beam could be created. It also had 24 transponders and associated amplifiers. That meant virtually any specific frequencies in the radio frequency spectrum could be tuned.

Jon's biggest concern was the security associated with control communications. The command channel did not appear to be encrypted so it was theoretically possible for someone else to hack that channel and control the satellite bus. He raised this issue with Bud.

Bud told him there were plans afloat to add encryption in future bus designs but the current bus firmware had no provision to store or run encryption, decryption algorithms. Instead it used a password. Jon was irritated with this revelation until Bud told him the password stored in the firmware onboard could be changed and it was 16 bytes long. Bud added the login to the bus used as the International Satellite Identification number.

The remodel was finally complete and after Marie and Katie had primed and painted the wallboard, the 'boys' were allowed to set up shop. The day of reckoning was a Saturday and Bud Levinson was there.

Unwilling to wait for any furniture or decorating, a folding table was used for the equipment that consisted of a multiplexer, a high quality oscilloscope and Jon's laptop computer. The cables from the dish ran to the multiplexer, a single cable ran from there, in and out of the oscilloscope to the serial, RS232, port on the computer. Everything was powered up and the command and control application was running on the computer. It was the moment of truth.

The exact coordinates for the satellite were typed in and the dish began to change its angle and azimuth. The oscilloscope was showing a wide swath of what appeared to be noise. Once the dish position was locked in place, the frequency of the command channel was selected. Almost immediately the oscilloscope showed a sharp spike in the center with very little noise on either side. As the frequency was so high no wave could be seen.

Bud said, "Well done. Before we try to log in, try altering the frequency slightly in each direction and let's see what happens. You might want to zoom the scope display."

They did and determined a slight decrease in frequency virtually eliminated any noise. Bud explained, "That's because there are certain interferences between here and there, like sun activity and the atmosphere."

The login menu option was selected and a common dialog box appeared. The satellite's ID and password had been provided. Jon typed them in. A previously red circle on the menu bar suddenly changed to green.

"You're now connected, congratulations!"

Marie and Katie clapped their hands; Jon and Karl exchanged a high five, and then shook Bud's hand.

Karl quipped, "Jon, see if Scotty is up there."

Bud said, "A few things before I leave. The password can only contain ASCII characters."

Jon interrupted, "Understood and with 256 available and a 16 character string, there's not enough time for all the computers to crack it in this century."

Bud agreed and went on, "As you know there are four antennas that can operate separately or in groups. In all probability they are currently working together covering a wide swath of North America. You should practice but consider creating your desired spot beams using the combination of two each.

"Last, you owe me nothing further; this has been fun and I'm available for any questions. Keep me in the loop."

And practice they did, hour upon hour for the next two days until they were ready for another test. Katie and Marie joined them to watch. Karl provided the coordinates for the Globalstar satellites they had been monitoring from Cabo. Jon adjusted the spot beam's focus for one of the two antennas they had combined.

He announced that the spot beam was a good approximation of the line of sight area from the Cabo receiver. He next adjusted the frequency range for the associated transponders to the correct Globalstar settings. They watched with great anticipation the oscilloscope screen. As before, they needed to make slight adjustments to minimize noise. The scope was quiet for over an hour. Karl was using the VPN to monitor their existing system. It was quiet also. They continued to wait until Marie left to fix and bring back some lunch. Halfway through roast beef sandwiches, both systems lit up. The scope was going crazy and Jon's laptop was recording the signal from the dish. The call lasted just under three minutes.

Jon asked Karl if there was a way to compare the two signals.

"No problem, let me download from Cabo, put it on a disk because you're not connected to the VPN and we can run the two streams through the spectrum analyzer." The tension began to build again. Karl returned with the disk, Jon inserted it and ran each signal through the analyzer. They were identical.

Karl whooped; Marie shook her head in amazement; Katie had a smug grin on her face. Jon turned and pumped his fist.

"Hey sweetheart, what's that look about?"

"Just thinking about the irony of it all; we buy a spy satellite with drug money to spy on druggies."

That comment brought a soft chuckle from everyone.

Jon said, "We have two more tests to run. Katie, isn't there a cell tower up the hill behind your condo?"

"Yes but quite aways up there."

"Doesn't matter, Karl would you find that tower in your database and give me the GPS coordinates? And Katie, would you mind calling Michael and ask him to stand by with his cell phone; oh, and ask him who is his carrier, please?"

Michael reported he used Verizon and was ready to go; a spot beam from the second antenna pair was focused on the right tower. Jon adjusted the beam width to the narrowest possible and then adjusted the frequency range to correspond to that which Verizon used.

"Someone, please call Michael back and have him first call the landline here. Talk about something for a minute then we will repeat the process having Michael call Katie's cell phone. OK? Let's go."

The oscilloscope had a built in speaker for use in the audible frequency range. Jon turned it on. A few minutes later he could hear Michael and Marie's voices although there were some noticeable delays. The call to Katie's cell came next. No discernable sound came from the scope. Both signal streams had been recorded.

"What did we just learn?" Marie asked.

Karl answered, "The call to the landline was not encrypted; the call between cell phones was. The fact the signals were not smooth like heard over the phone, means we either need to adjust the frequency band slightly or see if we can jack the gain on the satellite amplifiers. After all it's a long way up there and back."

Katie asked about the other satellite. Jon explained it was in a fixed position on the other side of the earth somewhere near the Urals. He confessed to not being sure how to use it from the western hemisphere, but had bought it knowing they would need to cover that hemisphere someday.

The next morning Marie and Stephanie were getting breakfast ready, and Jon and Karl were sipping their coffees in the living room when Katie came bounding down the stairs wearing a big smile.

"Good morning sweetheart."

"And a good morning it is, in fact: 'Hooray, hooray, it's the first of May, outdoor screwing starts today'!"

Jon choked, so did Stephanie from the kitchen amid the laughter, "Where did you come up with that one?"

"Oh, it's been around for a while; want to take a hike?"

"You mean like up the stairs?"

"Well maybe later, it's our wedding month and I'm happy. Can you tell? We need to get invitations out."

Marie and Stephanie brought out fruit cups, toasted bagels with raspberry cream cheese, and greeted Katie.

"Did I embarrass you?"

"Not at all," Marie responded grinning. Stephanie also grinned.

Katie commented, "You 'boys' have been up for quite a while. The clock read 5:36 when Jon got out of bed. Any problems or issues, I thought you pronounced the new system good to go?"

Karl responded first, "It's not yet a system, that's the problem. What we have created here is like the receiver part of our ground based systems. They have a back end; this one doesn't. That's what Jon and I have been discussing."

Jon added, "We can't just use the backend already in place and running. The architecture is essentially the same but the processing for the satellite receiver is different and the magnitude of the required data analysis and storage is very large.

"We were just talking about locations and all the backend equipment needed. Here, Cabo or Tucson doesn't really work, especially if we have to invoke the use of the second satellite. But before deciding that, we were wondering about the status in DC. It's been very quiet. Should we continue to work with the control group, DC would be an ideal base of operations as it's on the edge of viewing both hemispheres."

"Maybe it's time to contact Gates," Marie repeated a much earlier suggestion.

Katie responded, "I've been following the news every day. It's been ugly. As Randy predicted, National Security Advisor Clinton Prescott was targeted almost daily by certain media pundits known to be against the current administration. The White House press secretary pushed back against all the loud criticism by reminding whoever would listen that the country was under attack from terrorists and other enemies each and every day.

"I'll give him a call on Monday. Seems the 'boys' have some work to do this week and I'm so relaxed here away from the worries and spotlight, I don't want it to end any time soon," she added.

On Thursday Katie's email dinged with a message from Edgar Greeley. The Subject was 'Congratulations'. The message went on to say, 'Such a well-deserved honor. We are so proud of you'. Before the salutation, there was a link to a *New York Times* article and a comment that the committee had been trying to reach Katie through Greeley. She clicked on the link.

Unprecedented Dual Pulitzer Prize Awarded

Kate Ruiz of the Arizona Star has been awarded the 2006 Pulitzer Prize for Investigative Reporting and also a Special Citation Award for contributions to the well-being of the country.

She stopped reading and sat back in the chair, staring at the ceiling, trying to assimilate her feelings. Marie walked by and was almost ready to ask if she was alright when she read the headline on Katie's screen.

Marie broke Katie's stare with, "Wow!"

Katie looked intently at Marie and asked, "Am I totally crazy to be pleased and un-wanting of any further attention at the same time? Is any privacy a complete illusion?"

"Absolutely not. Stay put; I'm going to get Jon and Karl."

Together they bent over the laptop and read the complete article. It was incredibly effusive and complimentary. It closed by reporting the awards ceremony was to be in New York on May 23rd.

Jon leaned over and gave Katie a kiss on the neck and whispered, "I'm also incredibly proud of you. Better call Edgar."

Chapter 33

Katie called the *Arizona Star* and Beatrice, formerly referred to as the 'battle ax' answered. Before Katie could ask for Edgar Greely, Beatrice said, "All of us are so proud of you and how you have maintained your focus and integrity after all that has happened. Emails, messages, and other accolades have been pouring in. I've put them in a special folder named Pulitzer. I can send you a link, username and password to log in and view. Would that be alright?"

"Yes, please do and many thanks for your comments and efforts. Is Edgar available?"

"Yes, I'll put you through and you're most welcome."

Her former editor came on the line and she listened to him drone on for a number of minutes while congratulating her and referring to her career highlights at the *Star*.

When done, Katie responded, "You know I owe it all to you and your yelling and mentoring."

Greely snorted, "My Katie! The Pulitzer Committee is desperately trying to reach you and schedule the award presentation. Write down their phone number and contact person. Ready?"

She did and said a heartfelt thank you and good-bye.

AG Gates answered on the third ring. Katie greeted him with, "We've been feeling lonely and neglected. How are you doing?"

Gates chuckled and said, "Good to hear from you Katie. It's been quite an ordeal around here the past six weeks. Congratulations on your awards. They are well deserved. How are you and Jon?"

"We're good and thank you. I'm having a tough time dealing with more notoriety. Remember in our last meeting, or conversation was it, trying to make a plan to move forward? Well, our little group now has a plan and we're at a crossroads. We would like to know what your intentions might be, if any, for reassembling the control group."

"Interesting timing this call, in the ways of Washington the Inspector General and both Congressional committees announced their findings last week; some scapegoats have been identified. There have been three people fired and a number of resignations. The dust is essentially settled but it consumed most of my time as Clint recused himself because of conflict of interests. He came out unscathed."

Gates added with a snicker, "Too bad the house cleaning didn't go farther and out of the media spotlight."

"Sorry to be so blunt, but that didn't answer my question."

"Give me a day to consult with you know who and I'll give you an answer. Good to hear from you and keep up those great columns; I never miss one."

'We'll see' was the collective response after Katie related the call. Jon had completed the design of the new system's back end and identified exactly what existing code needed modification and what new code was required. Karl was nearly complete with his recommended new equipment list.

They were sitting outside enjoying a drink that evening. It was reasonably warm in Seattle.

Marie wondered out loud, "Suppose the control group is dead. Where do we go from there?"

Jon chose to answer, "It's first and foremost about our safety. With the new system we will be able to monitor Guzman's every movement and Zambada or any other threat. Now we will see them coming with no blind spots and with Michael's resources, eliminate any possible attack. Should we be able to shake Guzman to the bone again, status quo will return.

"And let me add, if the government doesn't want our assistance, I believe our consciences should be clear given what we've already provided."

Karl offered, "That's not wishful thinking, but I believe Gates and his group will come to the party. Then we really need to have our act together."

"I agree with Karl," Katie said.

Marie asked another question. "Is the control group the main reason why you think the capital is a good place for the new system? And if they don't choose to play, what about here or Tucson or even Cabo?"

"Basically that's the reason because if we're sharing intel and supporting operations, that's where the players are. I sure as hell don't want to live there and that's something Katie and I have yet to sort out. Also DC is on about the right latitude where we can see and control both satellite's positions while still located in this country."

"Wouldn't you or someone be required to monitor like we do now?"

"You're right about that. Remote monitoring only works as long as the equipment continues to function."

Karl interjected, "Jon and I have rolled around the idea of hiring Randy, Don and Charlie plus support staff."

Each pondered all the implications of the discussion they were having when Jon changed the subject. "I was reminded the other day when Michael called in that he was supposed to provide a bank or investment account so I could wire their retirement funds. He hasn't ever done that."

Karl spoke up, "And your use of the investment word reminds me of the ACT stock conversation I want to have."

"Let's have it now."

"As I said before in Cabo, ACT's market share keeps going down; new business is hard to come by; and the percentage of their revenues that comes from existing maintenance contracts keeps growing. You know that can't last.

"I get and read investor update reports from both my personal broker and from ACT's market maker all the time. A month ago, in a newsletter was some analyst's prediction that Bear Stearns would be the first to go, followed by Lehman Brothers. It seems much of their collateralized debt is almost worthless. It feels to me like the lead up to March 2000 when ACT stock lost 78% of its value. I'm ready to bail; I know you saw it coming before and shorted the stock last time, but as an insider I can't do that."

"I haven't paid much attention to my portfolio for a very long time. Just sort of knew it was there and used it for well-known purposes. I've always trusted your insight. Maybe it is time to bail as we're altering just about everything else in our lives."

Marie interjected, "Thanks Jon, Karl has been stressing over this for several months and I also trust his instincts. What do you see Katie?"

"I'm along for the ride on this, but trusting one's instincts is my mantra."

The next day, in another fit of conscience, Karl called his former boss and CEO of ACT. He explained his desire and need to sell his stock and wanted to alert the company because it would create downward pressure on the price. He also informed him to expect a communication from Jon Reynolds. He would not be returning as they were in the process of major lifestyle changes together. The CEO appreciated the heads-up and asked that Karl not put in any sell order until they could have another conversation within a week's time.

The conversation took place the next morning. Apparently the Board of Directors didn't want their market makers to learn an insider would be selling a large block of stock. Karl received an offer for the company to purchase his stock at that day's closing value and return it to Treasury Shares. Karl agreed and was told documents would be prepared and a closing date set.

After listening to what Karl had decided to do, Jon contacted his broker and placed a series of sell orders.

They met in the Oval Office again, but this time the President's Chief of Staff was not included. Gates provided a summary of Katie's call, confessing he had no idea what type of plan would be presented to the control group. The President was clearly intrigued. Prescott was quietly hoping to have his leash removed.

The President spoke up, "I don't see what we have to lose by listening to them. We've always been amazed at their abilities. To avoid having Clint followed by that gaggle of reporters out there, I want the meeting held here and have them transported in a vehicle where they can't be seen. Set it up."

Katie was sitting up in bed with her laptop open and typing into an Excel spreadsheet the names and known addresses of wedding guests. She had contacted the Pulitzer Committee, listened to another round of accolades, and agreed to be in New York for the presentation. Karl had been requested or summoned, depending on one's viewpoint, to a meeting at ACT by the CEO. Jon was writing code. Marie was doing laundry.

Katie's cell phone, lying on the comforter beside her, rang. It was AG Gates, "We have a green light to reconvene the control group. Both Clint and I are very pleased. It will be at the White House this time. What dates work for you?"

"Does May 15th work for you? That gives us enough time to prepare. And I need to be in New York on the 23rd."

"I'll confirm by email but let's plan on it and also let's discuss ahead of time your travel plans and housing."

"Your point is understood and taken. I'll be in touch and many thanks."

The four of them, Stephanie was in her room doing homework, sat around a fire. The early May weather had turned cool and wet. Marie asked if the meeting date would conflict with wedding plans.

"I really don't think so. Thanks to you it's very well organized and all that's needed is for Jon to go over the invitees. I want to get the invitations out in the next day or two."

"Why don't we just go to Vegas?" Jon answered sarcastically.

Marie was on him in a flash, "This woman deserves a wedding with all the purpose and trimmings. Don't even try to trivialize it. I'll have your head on a platter."

Jon could not find anywhere to hide, so he meekly replied, "Sorry, guilty as charged. In my only defense, it feels like Katie and I have been married almost forever."

Katie rose and stepped over to Jon, wrapping her arms around his neck and kissing him on the cheek. The others applauded.

Marie said in a calm voice, "I love you Jon and have always been in awe of your talents and value our friendship, but no apology for my comments. You deserved it."

Karl quickly changed the conversation, "We should be prepared to set up in DC as soon as possible and maybe even give our prospective employees a preview, what say?"

"Let me do some DC real estate research while Jon goes over the invite list," Katie offered.

Katie took to the Internet and searched Washington DC listings that might suffice given their criteria. She scored almost immediately by finding the eighth floor of an eleven story office building on New York Avenue available. The building had recently been completed and the owner was willing to do custom modifications of the floor to suit with a minimum three year lease. She provided her email address and requested complete information about the current floor layout and lease terms and conditions. The owner recognized Katie's name and became most effusive. Her email dinged less than ten minutes later.

Jon was feeling remiss when he contacted Don and Charlie. It was 2 AM in Seattle. He hadn't talked with them in many weeks and apologized after Don answered.

"No apology required," Don said, "let me grab Charlie. He's right outside and put it on speaker."

Don continued, "I just did the time zone arithmetic and you're up late; some special reason for the call?"

'Yes and no," Jon answered, "but first how are you guys doing?"

"Very quiet actually; the limited intel we've captured has come from the cell monitoring system, very little sat traffic probably because they're all dead. The Marines scored a major hit on the remaining Majeed clan. Truth be known, it's been somewhat boring since you left; our tour ends in just over a month and we don't know what will happen then."

"I have a question that comes out of left field. Would either or both of you consider private employment when your tour is over? It would be working with me."

"Doing what?" Charlie asked.

"Intelligence gathering, sorry but that's all I can say for now, and highly technical."

"Where can we buy you a beer when we're stateside?"

"Write down Katie's cell number. Stay safe and we'll talk again when you're home."

Jon tried not to disturb Katie as he crawled into bed. Awake or not, she rolled into him and flopped her arm over his chest. He was asleep in less than thirty seconds.

Karl's call to Randy produced the response, "I'm so ready to get the hell out of here; I'm almost willing to clean toilets." They agreed to meet in DC before the meeting.

Katie carefully went over the pictures and drawings of the DC property, and armed with Jon's and Karl's requirements, contacted the property manager for the eighth floor of the office building on New York Avenue. It was almost too perfect and after agreeing to a minimal buildout and lease terms, she finalized the deal. A deposit would be wired immediately.

In the midst of all the preparations, Katie received a call on her cell from Barbara Walters. They had exchanged emails after the Pulitzer awards were announced.

After a greeting, Barbara asked, "Understand you've agreed to be here for the awards ceremony. Can we schedule a guest visit on the show?"

"Sorry Barbara, not this time. We have nothing to add from our visit last fall, and in particular do not want to lie or disclose what we've been doing since. Both Jon and I would love to see you for lunch, maybe, when we're there. Privately I'll provide some insight."

"I'm intrigued as you might expect. Looking forward to seeing you two, all the best."

Chapter 34

It was mid-spring in Washington DC and the famous trees were budded out. The charter flight from Seattle used the Lear as no other planes were available. Karl accompanied Jon and Katie; Marie stayed behind because of Stephanie and school.

Their charter arrived at Andrews and quickly departed. They were greeted by a familiar Marine and driven to equally familiar base housing. Preston Gates called to briefly wish them a welcome and confirm the meeting date and time. Once settled in, Karl was housed next door, Katie began preparing sandwiches for dinner from the lightly stocked frig and dry foods available, while Jon pondered the details of his design for taking digital signal processing technology to the next level, and in parallel, wondered and worried if he and their team would be stepping into a cesspool of conflicting, politically motivated positions.

Karl joined them for a dinner consisting of roast beef sandwiches on toasted wheat bread, where the flavor came from a mustard, mayo, and garlic sauce. Jon continued his musings. Both Karl and Katie sensed what was happening and gave him needed space.

They had two free days to set up the DC base of operations before the scheduled meeting and laying out their plans. The office space on New York Avenue was complete and ready for full occupancy.

Katie decided to rent a non-descript Ford SUV so as not to rely on base transportation. The NSA spying stories continued to appear in the media along with a few almost outrageous editorial opinions. Katie never missed a deadline for her weekly column, but chose topics far away from national security issues.

They drove the rental to their new office space and examined it carefully, then pronounced it most satisfactory after deciding the future use for each room. Karl started to make notes regarding furnishings and other necessary infrastructure.

The members of the control group straggled in over a two hour period and were treated to light snacks while they waited. At 3 PM, Jon, Katie and Karl arrived in a black Lincoln with darkened windows. They exited the car out of sight and quickly entered the White House.

The President, his Chief of Staff who they remembered from the past August ceremony, and AG Gates stood in greeting. They were escorted to a room where the control group was assembled with an addition, the Vice-President. After brief introductions and greetings, the President opened the meeting. He was obviously in charge.

"Clint has briefed me on all the contributions you have made and the successful operations as a result. I extend my sincere appreciation to patriots such as you three."

Katie responded, "To that point, our thanks to you and the brave military personnel who conducted those operations. The upside has been a dramatic disruption in drug flow, Taliban combatants and enemy communications. The downside for us is we're back in Guzman's sights. Rest assured, somehow, somewhere, sometime we'll put an end to that.

"Now to the purpose of this meeting; Jon will be giving the main presentation but I have a number of blunt remarks to set the stage. It seems the spokesperson role falls on me."

The President interrupted her with a chuckle, "You usually are Katie and I read your column every week, congratulations."

"Thanks, each of you here should know our basic distrust of government and the recent NSA disclosures are just another example. We have chosen to trust this group and that trust has been clearly shown to be justified. I don't put blame for past NSA actions on anyone in this room; you preside over an unmanageable environment, but as Truman said, 'the buck stops here'.

"Jon, Karl and I, along with the other members of our team have struggled with defining our paths going forward. We have reached some contingent decisions and devised a proposed plan … hence this meeting, because all the contingencies involve you.

"Cutting to the chase, our proposed plan consists of creating and operating an intelligence gathering system that goes well beyond what we've currently been doing and also well beyond anything else in existence. Before anyone scoffs at those words, please remember the backgrounds and capabilities of both Jon and Karl.

"Before turning this over to Jon, I'm going to list a few of our ground rules. That probably sounds extremely arrogant considering whom I'm addressing, but so be it. Personal experience has taught us well.

"First, none of our team will ever have an employment relationship with any government agency; second, we mutually agree on every intelligence target. Next, no court upheld laws will be broken as a result of our activities and at the insistence of team member Sgt. Michael Ortiz, should a target be located on US soil, intelligence is immediately shared with local law enforcement; next, all intelligence gathered is first analyzed by this group before any outside dissemination or subsequent operations are planned or executed. Last, should one sentence about this group or our cooperation appear in the media, it's over immediately and the penalty clause in any agreement will be painful. Would you like Jon to proceed?"

After many glances back and forth between the assembled group, the President responded, "Of course you realize we can't agree to anything without hearing your plan, but I will say I don't find any of your 'ground rules' offensive. Please continue."

"Ready, Jon?"

"I want to echo Katie's thanks for your support. I also want to go beyond the cursory introduction and formally bring into this group, Karl Oberling. He is one of the founders of Advanced Cyber Technology, the world's leader in encryption and decryption technology. Until his recent retirement he was the Senior VP of Product Development. He is my former employer and a best friend. Please welcome him. On a related subject, Karl and I have sought and received independent legal advice concerning both our past relationships with that company, the plan we are now presenting and the laws Katie referred to. We are clean on all accounts.

"Before I continue, would you please bring us up to date on the operations in planning since the last time we spoke?"

Gates described Gwadar but was almost embarrassed to report nothing against Mullah Omar because of international politics.

"We're certain Pakistan is protecting Omar under the direction of President Musharraf. Matters are very unstable there right now and no raid can occur without his and ISI's cooperation unless we're prepared for the consequences. The President is not at this point."

"Thanks for that. We want to open this presentation with a somewhat rhetorical question; what would be your interest in renewed cooperation if I told you we are capable of intercepting any over the air communication, anywhere in the Northern Hemisphere?"

It was silent while the control group absorbed Jon's words and their implication. CIA Jim broke the silence, "What do you mean by 'over the air'?"

"All communications such as voice, data, and text except where each end is a landline; I might add there is one other limitation and that is we can only focus on up to eight somewhat narrow geographical areas at one time."

Jim asked another question, "And you're able to decrypt these communications?"

"Yes."

The President asked, "Will you explain how you can do this?"

"Not at the present time; there are a number of issues to work out first."

Before Jon could continue, NSA Prescott asked, "How does this relate to our existing intelligence gathering capabilities and NSA in particular. You certainly know what we've been dealing with there?"

"It doesn't relate. We're either complementary or a replacement in certain areas, but not integrated. Let's get to the issues; maybe that will clear up a few things.

"To begin, I'm going to use the word 'enemy' to describe any mutually agreed upon target for intelligence gathering, be they in war zones, drug trafficking, terrorist cells or wherever. Next we need to understand that to obtain intelligence information about a particular enemy or threat requires some prior knowledge that would define general geographical locations. Further, and to avoid sweeping up irrelevant and innocent communications, specific filtering criteria must exist.

"I won't be so presumptuous and try to tell this group about the laws pertaining to the ownership or privacy of communications. You know them better than us and I'm referring to longstanding laws before the Patriot Act such as the various flavors of the Telecommunications Act.

"In short, you provide the target along with supporting information and criteria; we mutually agree to focus our capability on that target and return the collected intelligence. The operational application or use of confirmed intelligence is decided by you alone.

"Consider this a public-private partnership. There are additional issues to address should we reach an agreement, but I'll field questions now."

Prescott asked, "Where do the systems in Afghanistan fit into your plan?"

"They don't because of their extreme limitations. However that location might be ideal for another purpose should our plan be accepted.

"As you are aware, there are two phone call intercept systems operating in Afghanistan, one directed towards low earth orbiting satellites that transmit calls between SAT phones within a narrow, line of sight range. The other uses a different antenna type and clones the operation of a cell tower, monitoring the GSM frequency range, but only as far as the next closest actual tower in Kandahar. We also have a SAT phone system operating in Mexico which is how we keep track of Guzman's calls. It should be pointed out that the enemies' suspect their communications have been compromised but don't know how or where. We have proof of this and are seeing a large drop in SAT phone calls and the content of the few is very circumspect."

Shelia spoke next, "Are you saying once the targets, as you call them, figure out what we can do, they'll change their methods and your intelligence gathering capabilities become useless?"

"Indulge me while I describe something you probably already know. In today's world desired intelligence about enemies is less dependent on human reports from so-called moles, spies and others. While 'lone wolves' do exist, the enemies posing national threats must communicate amongst themselves. They are geographically widespread. Even local terrorist cells need to communicate with either their sources of direction or sources of materials. And how do they communicate? Cell phones, SAT phones and the Internet; the old days of pay phones are about gone. And every type of cyberspace communications today is encrypted. Bottom line they must communicate and can only use existing technologies; they can run but they can't hide."

Karl summarized, "Our plan consists of the interception of enemy communications by any known airborne type, with certain needed capabilities provided by you as Jon explained earlier. We have the ability to do it and provide real time enemy intelligence and locations to this group for subsequent analysis and actions. The only caveat is we can't monitor trillions of calls each day; the specific targets must be identified along with approximate geography."

Jon said, "Before more questions, allow me to offer an example scenario. Suppose you or an ally suspects the leader of a mosque in London of planning a terrorist attack. We can intercept any and all communications either by sat, cell or wireless Internet within a reasonably wide distance from that mosque. One more example that's closer to home for us.

"Suppose we had obtained Guzman's cell phone and SIM number, we can already monitor his SAT phone; we could run a scan of selected towers and determine if it was turned on. Using one of several techniques, we could then, with reasonable accuracy, locate him and listen to any calls."

FBI James posed the next question, "We have been most impressed by your technology so far and its outstanding results. Am I hearing you right that this upgraded or new technology is not available or for sale?"

"That is correct. The gathered intelligence would be shared only with this group, and the underlying methods and technology not disclosed to anyone nor patented."

Randy asked, "Where do the NSA and our other intelligence agencies fit in, if at all?"

"The NSA is specifically excluded and the reason for that should be obvious. They have their own intelligence gathering methods that I'm sure you receive and use. Let them continue on their own. "

The Vice-President spoke for the first time, "You mentioned relevancy criteria. Who establishes the criteria and who decides what's relevant? Isn't this the same cesspool NSA stepped into?"

Katie answered those questions, "Sir, I understand your concerns under the circumstances. I will try to answer your questions. From my reading of the leaked documents, neither the NSA nor any intelligence department appears to have firm, established criteria. They only react to situations and swoop up enormous amounts of data, most totally irrelevant. The recent leak exposed the fact certain NSA personnel discovered sensitive and potentially inflammatory domestic information they attempted to utilize for personal gain.

"For a change we are suggesting this group, in concert with others under very controlled conditions, as each person here has a direct pipe into their respective departments, establish the target criteria and convince our team of its relevancy. Call us civilian oversight, and if that's too tough a pill to swallow, we can end this now.

"You decide the enemies or targets as we've labeled them; provide their suspected geographical areas; and request us to monitor and intercept communications that meet the mutually agreed upon relevancy criteria."

The Vice-President didn't follow up with any comment.

Shelia from Homeland Security asked the next question, "Understanding that any call you intercept must be decrypted, are you able to decrypt all calls and will you be sharing that technology with this group?"

Jon's answer was, "Yes and no. There are two primary SAT phone providers, Globalstar and Iridium. We can decrypt both and are currently doing so. There are three cell phone vendors that control 95% of the market: Nokia, Motorola and Blackberry. We can decrypt all three, and should another vendor appear, we will develop a decryption method for them. To your second question, absolutely not and it has nothing to do with the integrity of this group."

AG Gates asked, "Please describe the agreement between us you referred to."

Jon glanced over at Katie and nodded his head.

She looked directly at Gates and said, "I am the Managing Member of a US based LLC that we have formed. It owns all the equipment and underlying proprietary technologies necessary to perform as advertised. In addition it will manage and operate the actual intelligence gathering. We are prepared to establish a headquarters here in Washington DC.

"At the risk of offending, we have certain criteria of our own. This administration will end next January and right now nobody can predict who will be in office. Our plan is not short term and we expect this control group be made permanent.

"We have a pretty good idea about the morass of agencies feeding Mr. Prescott. In our viewpoint only one has a good long term track record and that's the Defense Intelligence Agency."

One star Gus looked up in surprise.

Katie continued, "We propose to have a contract with the DIA that generically describes a consulting relationship for a small annual fee. That contract and any and all materials we may exchange must be classified at the highest level. The ball is now in your hands."

"Hard bargain," the President said, and then added, "But not without precedent. I'm guessing there are many layers of detail you haven't presented, but we get the basics. Let us discuss. Preston will call you very soon, one way or the other."

Chapter 35

They left the meeting and the White House with mixed feelings. They were uncertain what decisions were forthcoming and wondered if they had spent a lot of time, effort, and money for nothing. The control group had remained behind.

Arriving back at the base they decided to walk over to the officer's club for a drink and some snacks. It had been agreed to stay on the base until a decision was reached to avoid any accidental recognition that would end up in the press. A conference call with Marie, Michael and Corinna was planned for that evening.

The President requested comments and opinions from the group. The Vice-President spoke first, "I have another meeting in fifteen minutes so I'll go first. Needless to say Ms. Ruiz is a most distracting presence, I probably shouldn't have said that but you know what I mean and it's a compliment. The issue I believe, beyond any potential legal problems, is trust. While I haven't been directly involved, such as you Preston, from everything I have learned, they have earned our unwavering trust and confidence. The fact they're choosing to trust us, at least to a certain degree, is remarkable.

"I must confess a strong desire to add their capability to our arsenal, God knows we need it. Unless someone here can point out a good reason for turning down their offer, I say go for it."

Gates spoke next, "I have worked with them for almost a year and simply marvel at their abilities and trustworthiness. I also recommend we proceed."

The President faced Gates and said, "Preston, burn the midnight oil at Justice and give us a brief on any possible legal ramifications. Anyone else want to comment?"

Shelia described her primary concern, "For me it's more about how to manage the criteria they're requiring plus the intel we might receive. I don't want to complain but you might be aware of all the pressure I get to disclose this group's activities and my superior's desire to micromanage my participation."

CIA Jim expressed similar concerns, adding, "You have it easy over there, General, maybe that's why they picked you."

FBI James stated his primary concern was the potential utilization of the system for domestic targets on his agency's list.

With no other comments the President said, "If we proceed, Clint will manage any turf wars and I will have his back. The authority of this group and its members will be authorized by an Executive Order. Preston, let's get to work."

After the conference call, Katie came out of the shower in a robe and announced, "Not bad and a lot better than the other base, great water pressure. What are you up to?"

She came up to Jon, seated in an uncomfortable chair, and before he could answer found he was nose to the waist tie of her robe. Then the robe fell to the floor. He stood and was led to the small bedroom, forgetting about everything previously on his mind. After slow and gentle lovemaking, Jon slept for almost nine hours straight.

Katie was up the next morning preparing coffee when her cell phone rang. She finished pouring water into the coffee maker and pressed the brew button before answering. It was someone from the gate at the base where they were staying.

The caller said, "Someone named Randy Allison with NSA credentials is here looking for Mr. Oberling. We didn't have a contact number for him. Should we direct him to you?"

She responded yes and roused Jon, then went next door and knocked on Karl's door. She saw Randy drive up and waved, then pointed towards Karl standing in his doorway before returning to her own. It was over an hour later when Karl and Randy came over. Jon was up, showered, dressed and feeling terrific for a change.

Karl broached the subject, "Randy wanted to have an offline conversation after yesterday. I reiterated my previous offer. He can't predict what decisions will be made. But should we put together a relationship, he would like our input to the offer as to how and when, along with any mechanics we're willing to share. I think it OK and fair to have this discussion."

"So do I," Jon responded.

Katie said, "We are anticipating a positive response and have taken a lease on an entire floor of a DC office building. We might have jumped the gun, but I don't think so. Karl, did you tell Randy about our similar offer to Don and Charlie?"

"Yes."

Jon picked it up from there, "Once we have a contract in place that space will be set up and occupied very quickly. Initially we would require your assistance for that set up and once fully operational participate in its management and operations. You, in particular, would be the onsite contact and interface with the control group. Don and Charlie would be more involved with operations, should they choose to join us. Other staffing needs are yet to be determined."

Katie added, "If you are looking for an organization chart, we're all peers with particular areas of responsibility."

"Except for you, boss lady," Karl quipped.

"Seriously, you would have an employment contract with our LLC and salary, bonus opportunity and full benefits. About your request for mechanics, we always take the frontal approach and disclose before any contract is signed. Jon can do that or I will."

Randy thought about it for a few moments, "I'm busting with curiosity about your new system, but will restrain myself. I trust you people so I'm on board."

"Terrific, don't you want to know how much?"

"You mean salary? I'm a GS18. Keep it a secret for now. Let's see what happens next."

They shook hands and Randy left. Katie kicked the 'boys' next door while she worked on her next two columns. They repeated their trip to the officer's club that evening.

In almost a repeat of the previous morning, except Katie was this time pouring her first cup when her cell phone rang. It was Preston Gates.

"Do you want to make a deal?" he greeted in a serious tone of voice.

Katie laughed and answered, "Are you watching TV or auditioning as a game show host?"

They both laughed and then Jon appeared, not quite awake wondering about all the noise. Katie waved him aside.

"Of course, what do you have in mind? And which door do I get to choose? Don't forget I'm older than you think."

"3 PM today, same place, will that work?"

"We'll be there, same transportation arrangements?"

"Yes, see you then."

The regular group was already assembled. The President and The Vice-President were absent and had been replaced by a stern faced, middle aged woman named Margaret, who was introduced as a lawyer from the Justice Department with 25 years' experience.

AG Gates opened the meeting this time, "We have conditionally decided to accept your offer and proceed, but first I would like Margaret to present our legal analysis and positions. Margaret?"

Jon, Karl, and Katie showed no emotions.

"Thank you Mr. Gates and it's with great pleasure to meet the three of you. As you can imagine we walk a tightrope in our efforts to protect this country, its people, and institutions. External and internal threats abound. I am told you have received legal advice associated with your proposal and planned activities. At the direction of Mr. Gates we have performed our own analysis which I am here to summarize.

"There are four applicable, broad statutes that provide the legal criteria for the relationship you propose. They are the Espionage Act, a 90 year old law, still on the books and amended and addressed by the courts many times since. To quote one section:

'... *This outlaws the disclosure of any information, documents or other articles relating to security or intelligence*'

"Next is the Foreign Intelligence Surveillance Act of 1978, also subsequently amended and court interpreted. It provides the legal protection for your activities and any warrants or orders by the FISA Court must be properly presented and documented.

"Next is the Telecommunications Act of 1996 which actually is a benefit to us, as it removes any and all barriers to private companies or citizens engaging in that business area, and regarding privacy issues, is amazingly vague. Courts have determined that law does not convey any ownership or special privacy requirements to communications, only the Fourth Amendment does that. And last, of course is the Patriot Act.

"As it pertains to a contract between us there will be two requirements. The first will be a clause that specifically includes the Espionage Act and the consequences of violation. Every member of your organization will be required to sign.

"The other requirement is the assignment of a Justice lawyer to act as a compliance officer for this group; insuring any and all activity complies with existing law. A signoff will be required. Mr. Gates?"

"Thank you Margaret. It should come as no surprise that Margaret would be the first to fill that role."

"I actually think those requirements are terrific," Katie responded.

"That's very good. One of the concerns expressed by people who don't know you as we do was about your public persona and continued media presence."

"I have already promised there's no story behind any of this. The public attention drives us crazy. Maybe we should wear Jason masks and a toy chain saw in public. That would sure keep people at bay."

Everyone had a good laugh.

Gates continued, "There are some organizational changes to be made. As you so aptly pointed out this administration is on short time. The President is drafting an Executive Order formally creating this group and associated contractual cooperation. Gus has accepted to be the Lead for the group and will also join Clint's Security Council. Clint will fall back to an advisory role unless something occurs requiring his direct involvement. That also applies to me, however Margaret will be reporting directly.

"You mentioned a small annual fee. What's small?"

Jon answered, "We will be establishing a facility here. We anticipate having up to six or maybe seven employees. The annual cost of maintaining that facility and its overhead will be the fee. I will provide a full accounting. There will not be any other costs or charges, including to members of the LLC or its equipment and property. Best estimate is a million and a half per year."

"That's incredibly generous. Are you sure?"

Karl answered, "Yes, we're sure and I have a disclosure to make, it's about…"

"Randy," Gates interrupted, "we were informed yesterday and give our approval. He decided to come forth and we understand his responsibilities will include being our direct contact."

Jon answered, "Yes, and I have a further disclosure to make. General, would you consider putting both Don and Charlie on active reserve and allow them to join us?"

One Star Gus thought about it for less than a minute, "You people are amazing; do they know anything about this?"

"Next to nothing, only that an offer might be coming when their tour is finished; for us, the sooner the better should you agree."

"What about the systems they're monitoring?"

"Once the facility here is operational, the first demonstration will be that geography. Then you should be able to decommission those and bring them home."

"I can support that," Gus said.

Margaret added, "Each of your employees will be required to sign appropriate security documents."

"No problem," Katie said.

AG Gates closed the meeting with, "Gus will be your point on the contract. Margaret will prepare the additional documents; Clint and I will watch it all unfold from the sideline and continue to marvel at this outstanding collection of talent and cooperation. Thank you to everyone and let success be our future."

They departed in small groups. While they waited for the Lincoln to arrive, Jon, Katie and Karl wandered along an ornate hallway looking at and admiring the artwork and historical depictions. Randy found them and was about to say thanks, when Katie gave him a brief hug and said, "Welcome to the team."

"This is such a great opportunity for me, thank you. Would you like to come for dinner tomorrow and get off that base? You will meet my family, and they sure want to meet you."

They accepted and were told directions would be emailed.

Karl took the rental car and went shopping. After being dropped off at the new office building, Jon and Katie met with the property manager to discuss the build-out.

What they wanted were two small, single bedroom apartments with kitchen and bathroom facilities as far away as possible from the main office areas. The rest of the floor and its current layout required no modifications. The next request was to have their own provided antenna mounted on the roof with cables run to a yet to be designated location on their floor. Their last request, or rather requirement, was no building directory listing for the eighth floor and some type of modification to the elevator's controls to only allow access to their floor by approval, key, code or whatever would work. He promised to get on it immediately.

Jon and Katie wandered around the large empty floor, lost in their own thoughts, when Karl returned.

"I was just in the most incredible office furniture store ever. I guess that should be expected here in DC. This place will be totally transformed tomorrow. I called Marie and she will place all the equipment orders on our list with expedited shipping. She was ecstatic at the news. We should go out for a celebratory dinner. I know a great little restaurant in Georgetown from prior visits. What say?"

Katie pulled the 'boys' together and they had a long group hug.

The restaurant was small, warm and specialized in Italian cuisine. The décor matched the menu. Karl had dropped them off in front then sought a parking space. A formally dressed maître d' greeted them and asked if they had reservations.

Jon answered, "Sorry but no, your restaurant was recommended and we thought we would try."

Whether it was because of seeming familiarity or the belief he was greeting some Washington power couple, the maître d' decided they would be accommodated.

Jon added, "There's a third person joining us, currently looking for parking."

Karl found them in a semicircular elevated booth towards the back. He slid in across from Jon and Katie who were close together. The wine was an excellent Tuscan Chianti, the food and service impeccable. Katie was considering a dessert order when she looked up from the menu realizing someone besides the waiter was standing next to the booth.

"Can we help you?" she asked.

Just then a camera flashed and Jon demanded an explanation resisting the temptation to grab the short man facing them.

"Sorry for the interruption, but it isn't often I run across such famous and attractive people. I'm a big admirer." He quickly left but the entire restaurant's attention was focused in their direction.

The maître d' came over, now knowing who they were, and apologized for the intrusion.

Jon told him thanks and asked for the check offering his credit card.

As they waited, an attractive woman around thirty approached their booth and placed a card on the table.

"My name is Rhea and I work for the Huffington Post. Welcome back to Washington. I covered your Medal Ceremony last August and we publish your columns. Should you ever grant an interview, please call me. I won't intrude any further."

Katie smiled at her politeness and said, "I'll give it some consideration. We love this city, outside of its sometimes ugly politics of course, and visit often. Have a good evening."

As they walked down the sidewalk towards where Karl had found parking, Katie commented, "I can hardly wait to see what shows up tomorrow morning. Better get used to it; can't get mad about what you can't control, but I sure know the topic of my next column. Are you still mentally prepared for our trip to New York?" Jon only nodded.

And sure enough after an Internet search their picture had been published several times with various captions. The most amusing one announced 'Reclusive Couple Spotted at a Romantic Dinner in Georgetown'. Karl was not shown because of the angle and his seating position. He feigned disappointment at not being featured. Katie retrieved the card Rhea left on the table and brought up that website. She had been quoted verbatim but with a slightly different picture. She wondered how that had been taken.

Jon added to the conversation that his last thought before falling asleep was a visual from some old movie where the flash attachment and bulb was smashed and the film ripped out from the camera. Katie laughed and said she remembered the scene but not the movie.

It was back to the new headquarters' office to continue setup and receive Karl's purchases. They were scheduled to catch the train to New York early Thursday morning and there were still many tasks to complete.

The property manager was found directing a contractor about their priority, the apartments. He informed Katie the elevator company was looking into their request and inquired if she wanted to contract with their current cleaning company. She told him no.

The furniture began to arrive in an almost never ending stream of trucks. They worried if it could all be completed before the dinner engagement with Randy. It was, with Karl playing a strange combination of traffic cop and decorator.

Randy lived in Tenleytown, a family oriented suburban area. The directions given were good and they arrived on time. The house was an older, well maintained, light brown two story with a gabled roof. The attached garage was obviously not part of the original. Randy heard them pull into his driveway, greeted them at the door and then took their coats in the small foyer. Randy's wife, a somewhat chubby, curly haired blond, was introduced. Her name was June.

She smiled warmly and said, "Just like the month and I'm so glad you're here and to meet you."

Katie reached out a hand and thought, cute and bubbly about the opposite of serious Randy.

"Please come in and meet our adorable monsters."

They were introduced to Tyler, age 12, Travis, age 9, Tommy, age 6 and Tucker, age 4, all towheads resembling their mother and very polite.

Dinner was a delicious crispy fried chicken with mashed potatoes and country gravy. June offered and served a complimentary white wine. It was consumed in its entirety. The four boys had disappeared after dinner and the five adults sat out on a small rear patio.

June commented, "I haven't seen Randy this happy or excited in a very long time."

Karl said, "A most welcome addition to the team. Actually we've known each other for what, about eight years now?" He went on to describe the other members of the team and assured June she would meet them all and experience an interesting collection of personalities, albeit totally compatible.

Jon asked Randy if he could find the time tomorrow to visit the new facility. Randy asked if it could be in a day or two because he was involved in exit interviews and some other discussions. Katie wrote down the address, stating Karl would be there the rest of the week but they were leaving on the 23rd.

On their way out, Katie took June aside and after thanking her and complimenting again the delicious dinner said, "I know Randy is excited but be prepared to enter a very different world from what I believe you've experienced in government employment. Please know you are most welcome."

June smiled, almost knowingly, and squeezed Katie's handshake.

Randy showed up in the building lobby mid-afternoon on Tuesday. He could not figure out how to get to the eighth floor so he called Karl's cell phone. Karl descended to the lobby and showed Randy the procedure. The elevator company had modified their system. When the button for the eighth floor was selected, the small LED emergency screen displayed a message asking for a code to be entered. Using the floor buttons, Karl entered 3, 6, 1, and 9. The doors closed, the elevator rose and opened on floor 8.

Randy could hear construction activity behind him and followed Karl into the main office area. He was beyond surprised and it continued as Karl took him on a tour. There was a lobby area with a large reception counter. Opposite was a conference room furnished with a long table and too many chairs to count. Offices with desks, chairs and cabinets continued on one side with three large empty rooms across the hallway. Karl explained they were for the actual system equipment, just beginning to arrive.

Jon and Katie emerged from the construction area beyond the elevators and greeted Randy. They chose to inaugurate the conference room.

Katie had been tasked with obtaining a boilerplate employment contract and edited it using Karl's that he provided from ACT. She also contracted with a local company for payroll services. Complaining that all these business tasks were outside her skill codes, Jon decided not to comment, but Karl reminded her she had been unanimously elected Managing Member and general spokesperson, aka mouthpiece, and business tasks fall into that category. Katie had nowhere to go after that.

She slid a small sheaf of papers across the table to Randy. He read them with a variety of facial expressions. His largest surprise was a salary of $250,000 per year, over double what he had been receiving as a GS18. The other benefits were equally generous.

Karl said, "Take it home and discuss with June. Let us know if we missed anything. We'll have to add whatever Margaret or any of the others require. What's your status at NSA?"

"The exit interviews are over and I will have completed organizing all my files for handoff tomorrow. We agreed to a one year support and consulting arrangement. When do you want me to start here?"

"The next time you show up," answered Karl.

It took several more days for all the equipment ordered to be received and unpackaged. The build-out contractor was nearly finished when the antenna arrived. The one on Karl's roof in Seattle would stay in place. The contractor experienced several difficulties routing the cables but finally completed the job. Karl had opted for a cluster of rack servers from IBM. It would provide more than enough computing power. Their technicians arrived along with the computers. Karl showed them the location and a day later the installation was complete. Karl sent the techs away without running any tests. They were more than surprised. The disk storage racks came next. The total was almost a petabyte.

After being shown the rote procedure for removing factory installed software, then installing their own clean operating systems and their application software, Katie had every workstation up and running in short order.

They sat back in the comfortable conference room chairs totally exhausted from all their efforts. Karl reported he could tell from Marie's voice she was becoming frustrated.

Katie stated, "Maybe we bit off more than we can chew this month. We'll be gone for the next few days, our wedding is eight days away and if there are no objections, I'm ready to check out. Let's hope we can finish all the tasks and be at the altar on time."

Her comment hit Jon hard; Karl decided it was time to leave the room.

Jon was about to speak when Katie placed a finger on his lips, "Don't say anything; no apologies required. I see almost everything. Finish here, be on time, say 'I do' and love me forever."

He scooped her up onto his lap and they held each other for a very long time.

Chapter 36

They caught the 6 AM train from Union Station using a cab from the base so Karl could stay behind and complete his set-up tasks without time wasted transporting them. They promised to meet again in Tucson on Memorial Day together with Marie and Stephanie. They sought another cab in New York after their late 9:15 AM arrival. Katie had arranged for them to meet Barbara Walters for brunch at a 42nd Street café near her studio. Somehow the paparazzi had anticipated their arrival for the Pulitzer Prize awards as a number of pictures were snapped while Jon was hailing the taxi.

Barbara Walters was already seated when they arrived. She stood and greeted them warmly. Over perusing the menu, Jon offered, "This was a great idea. I'm actually hungry and the house omelet looks terrific."

Katie knew Walters was anticipating some type of disclosure – news folks are all alike, and she had been thinking about what to say during their train trip. She also knew she owed Walters, not for the media exposure, but the warning that essentially saved their lives.

She looked up at the standing waiter and ordered OJ and a cinnamon roll. When it arrived, and before Barbara could ask any questions, Katie began.

"I know I've said it before but you have our undying gratitude for the warning last year. Without it Jon and I would have been oblivious to the threats we faced. Both Jon and I believe in your integrity that anything I say stays between us. I owe you an explanation for my refusal for us to visit your show again.

"Last September, about the time we last appeared on your program, I received an email threat from El Chapo that said 'YOU ARE ALL DEAD'. I'm reasonably certain you are aware of the Seattle policemen murders and the attempt on Sgt. Ortiz. There have been other attempts."

Barbara nodded and Katie continued.

"After essentially hiding out for four months, we, including the members of our group or team, have been in a constant effort to ensure our security and our continued lives. This has led us across the world and we have created systems to protect us. We have taken this battle to El Chapo himself, as he gives the orders for our deaths. At the present time we feel relatively safe. I hope that gives you some small understanding."

Barbara Walters picked at her fruit cup, scrambled eggs and cheese, obviously absorbing Katie's words. At length she looked up and responded.

"Here I am, relatively safe in New York. At least I don't worry about or dwell on my personal safety. I can only imagine what you are dealing with in your world. Rest assured anything disclosed is only between us. You already know my role and position in the media, but as a fellow human being, I also know information extracted and disclosed for audience rating reasons must be filtered to protect the sources and their unique situations. You have no problem with me.

"I have seen the several media reports of your comings and goings, in particular in DC. With my trained nose, I'm certain there's much more that you're unwilling or unable to disclose. I'm there if and when I can be helpful."

Jon said, "Ms. Walters, you are truly an amazing person and have my respect."

"Barbara, please, Dr. Reynolds."

They all chuckled and departed on good terms.

The Pulitzer Prizes are administered by Columbia University in New York as they have been since 1917 and the awards ceremony, in twenty-one categories, is held at the university at 2 PM with a light lunch served.

Katie and Jon arrived with a half hour to spare and wandered around the amazing architecture of the university until it was time. Katie was immediately recognized and descended upon by several hosts. They were escorted to front row seats. The ceremony began on time with a medium length dissertation about the origin and purpose of the prizes including a brief biography of Joseph Pulitzer.

As the ceremony dragged on and various recipients accepted the prize in their category, and then made a few perfunctory remarks, it became obvious to Jon that the Committee had decided to leave Katie's unprecedented dual award to the last.

The Pulitzer Committee member previously at the podium, presenting awards, changed. The person now standing there was recognizable from the event program to be the Committee Chairman. He raised his hand for silence and began in a very surprising lower alto voice.

"Ladies and gentlemen, we now come to an unprecedented decision by the Committee – a dual award.

"The category is Investigative Reporting. Measured by our over 85 year standards for research, accuracy, documentation and relevance criteria, we have never before encountered anything to equal the reporting of Ms. Kate Ruiz. As such she has been awarded the Pulitzer Prize in this category. Please stand Ms. Ruiz."

She did and received a loud ovation while continuing to face the podium.

The Committee Chairman continued, quieting the audience.

"Before we invite Ms. Ruiz to the podium to receive her award, the Committee decided, during extensive deliberations, to add a second, Special Citation Award that has only been given four previous times since 1917. The description for this Special Citation reads: For journalistic content providing a significant contribution to the well-being of this country.

"She addressed social issues of extreme importance and articulately defined methods to alleviate these issues. Please welcome a truly remarkable journalist, Ms. Kate Ruiz."

Katie seemed to be unfocused and Jon had to nudge her several times and help her to the steps up to the podium. He whispered an order to 'wake up' in her ear.

Katie released Jon's guiding hand and mounted the steps normally. Her eyes were clear and focused as she accepted the outstretched hand of the Committee Chairman. All the while there was a standing ovation.

The two gold medallions were handed to her. She nodded a thank you and placed them on the lectern. Then she faced the audience and waited for them to sit and become silent. An uncomfortable silence ensued.

Katie spoke at last, her voice strong and clear, "Thank you Pulitzer Committee for these prizes. And thank you audience for your applause. I am honored and humbled.

"No investigative reporter operates alone or in a vacuum. When an investigative reporter's output is based on the collaborative efforts of many others he/she can only accept an award in the context of their contributions. For the Investigative Reporting prize, that is my situation. I was one part of a very special team. The other members are well known. I accept this prize conditionally on their behalf. The prize will be displayed at the *Arizona Star* with full credits given and dedicated to the memory of two murdered team members.

"Why does an investigative reporter expend the effort and face potential danger to complete such a piece? Speaking for myself, it is to right a wrong, expose corruption, or any other travesty innocent people must endure. The most important motive should be that their efforts will affect a positive change.

"With no disrespect, I cannot accept the Special Citation Award. Referring to the prize's description, there has been no real change as a result of our efforts. Who will benefit from all the illegal drug money being seized? No person I know. Almost unabated, drugs of all kinds continue to flow into this country, bought and used by disaffected citizens, those only looking to relieve the pain and depression of their lives, or to create a surreal environment where the view of their own self-importance is chemically realized.

"Bottom line, Committee and audience, with no real change from any investigative reporting, no deserved award. I don't stand here to be awarded when nothing was achieved. Thank you."

There was minimal applause as Katie left the lectern, the Special Citation medal still there, walked down the stairs, reached out her hand and pulled Jon to his feet. Together they walked out of the university and hailed a cab, to catch their flight to Tucson and their wedding.

Katie was back in the news, from coast to coast, over the Memorial Day weekend. Her words were reported verbatim in nearly all the syndicated outlets of her weekly column. Editorials were many, most positive and supportive, some not. She was a topic of the Sunday talking heads. On Monday a new focus occurred, that in the celebrity media.

Somehow it leaked, how and where they never knew, but it was originated by People Magazine and somehow spread across the Web: news of their wedding. They suspected at least one of their invitations had fallen into the hands of a media outlet.

They had chosen an outdoor wedding and the venue was the Saguaro National Park, more particularly a tended and landscaped area near the trailhead parking lot they had used before. A Justice of the Peace had been hired to conduct the civil ceremony and they had received approval to erect temporary bleachers capable of seating the nearly 100 invited guests.

Karl had finished up in Washington DC with three days to spare. After a stop in Seattle to pick up Marie and Stephanie, the Oberling family was in the newly refurbished guest condo, two units from Katie's. Jon secretly visited a Men's Warehouse and selected his wedding suit. Katie was equally secretive about her chosen dress. Upon their arrival in Tucson, he tried to put a move on Katie but was rebuffed for an understandable reason.

The morning of the wedding, scheduled for 2 PM, found Katie, Marie, Stephanie and Corinna leaving to have their hair done. Jon was pacing around lost in memories from his past. Karl and Michael stayed out of sight; some friends, Jon thought.

Finally Karl knocked on the door, opened it part way and asked, "Time to go; are you ready Jon?"

Jon opened the door fully, "Man do you clean up nice."

Jon's new suit was dark blue, the white tuxedo type shirt starched and pressed and the tie was a patterned red, white and blue. "Yeah, I guess," he answered, "Where's Michael?"

From the garage, "Right here, let's get a move on."

They arrived a half hour early. The temporary bleachers were nearly full with Katie's uncle Hector and the extended family, Lucinda and Ricardo and much of the staff from the *Arizona Star*. Sr. Valenzuela and several members of his family had made the trip to attend. Edgar Greeley had a front row seat and to Jon's utter amazement, Preston Gates was seated alongside him with Wes Davis, the Pima County DA, on the other side. Jon walked up to shake their hands.

Gates said, "I wouldn't have missed this for a thousand crises. Congratulations Jon, I can hardly wait to kiss the bride."

With only minutes to spare the four women arrived. When Jon saw Katie step out of the back seat, he was dazzled. The dress she had chosen was a champagne color with a tight fitting bodice, low in the back and a flowing floor length skirt. Her hair, nearly grown back from the December cutting, was styled by pulling one side back from her face into a cluster and the other side many wisps and tendrils. She was a beauty to behold.

Michael walked Katie down the aisle; Marie and Karl had preceded them with Stephanie close behind. Michael then stood beside best man Karl and the ceremony began. They had not planned any music, Hector took care of that. Just as Jon and Katie completed their vows, kissed and turned to accept the audience's applause, the sound of the Mariachi Band playing La Paloma was heard. As they walked together, now husband and wife, Katie was swaying to the music with an enormous smile.

Anticipating a media presence, Michael had hired perimeter security for the wedding. It turned out to be a good decision because about two dozen media types, including photographers, were kept at bay, but considering the venue not that far away.

Hector and the crew from El Matador catered the reception. It was a feast to behold. Jon could hardly wait to get the receiving line over with and chow down. Katie received so many hugs and kisses Jon was almost jealous, but in his reticent way smiled, said thanks and shook hands. Katie's embrace with Edgar Greeley, Sr. Valenzuela and Preston Gates held up the line.

If any of the wedding guests thought Jon and Katie different or a bit crazy, it was confirmed after the reception. They both went into the nearby public restrooms and emerged wearing hiking boots, shorts and t-shirt tops. Their wedding clothes were neatly hung in garment bags and placed in the care of Marie.

Donning their medium size backpacks, they bade their guests goodbye and headed up the trail for about two miles to set up an overnight camp for their honeymoon.

Chapter 37

In the wind down from the wedding the following weekend, two new alignments formed: the 'boys' and the 'girls'. With the school year essentially over Marie and Stephanie decided to stay awhile longer. Together with Corinna, Lucinda and Katie they became the 'girls' and made their own plans.

The 'boys' now including Michael and Ricardo, went back to work and a charter flight was arranged to return all of them to Washington DC for completion and testing of the new facility. They arrived on a Monday, Randy's first official day on the job. It was June 4th and the 8th floor apartments were complete. They flipped a coin to decide which apartment and who got the bedroom and who got the sofa bed. Michael confessed to Ricardo that he snored.

Karl identified each unopened box and stated all equipment had been delivered and accounted for. He then found his way to the roof and was satisfied with the antenna installation. Michael had been busy defining and procuring the needed security equipment and services. By the end of Wednesday the facility was ready for testing. With Randy, Michael and Ricardo watching, Jon and Karl repeated and explained each step they had done in Seattle. Within an hour they pronounced the system ready for action. Only the contract details remained unsettled.

Randy commented, "Government agencies live in their own time zones."

Karl came back with, "I know, that's only too true."

Jon said, "Now that we're ready here I had a sudden brain flash wondering what the women are doing."

Karl responded, "I don't think we want to know."

"God help us all," Michael added.

Ricardo just snickered.

Jon's fertile brain was not ready to shut down. "In the morning Karl, please give Bud a call. He alluded to a way to control and include the second satellite. We need to get it operational."

Realizing they had not fully briefed Randy, Jon mentally decided to settle all outstanding contract issues the next day.

Karl agreed to the request. They called it a night.

After reading all the media accounts of the Reynolds, Ruiz wedding, El Chapo was on the phone with his Mexico City employee giving the following instructions:

'Contact Panama, require meeting at same location as before, he'll know where. Courier back earliest date and time with Jesús.'

The system in Cabo duly processed this call and inserted the call log into the database. Guzman was being as careful and circumspect as possible with his needed communications. He was certain the requested meeting could be held with no risk.

Both the DIA General and Margaret reported the contract documents would be ready for review by the end of day. They offered to be available over the weekend should there be any questions or concerns. The only item they were missing was the registration information for the LLC. Jon promised to email that within an hour.

While they waited for a call back from Bud, Randy was given a full briefing and shortly thereafter Bud returned the call.

Jon posed the question regarding the second satellite. Bud reminded them of their previous conversation by explaining again the satellites could not directly communicate with each other because they were essentially on opposite sides of the earth over the equator in the GEO band, and a Western Hemisphere antenna could not be focused on the eastern position.

"What is needed is another ground based antenna such as you have there with line of sight. Control of that antenna can be accomplished with any link or connection, such as the Internet. I would recommend a duplicate of what we set up in Karl's home, with a computer connected to that antenna, running the same command application over the link.

"The downside of using the Internet is the effect of latency on the captured signal or data stream, but there are TCP boost and acceleration methods available to minimize those problems. Does that help?"

Jon responded, "Absolutely and I'm very familiar with those techniques. It was really the ground antenna location I was foolishly hoping to overcome, but I think we have a solution for that. Thanks Bud, bye for now.

"Karl, better order another antenna, cables, and scope. Randy, how would I go about getting authorization to use the communication link you set up from here to the base in Afghanistan?"

"Do you remember Gloria over at DIA?"

"Of course,"

"She was the one who actually set it up as it was going to a military facility. Shouldn't be a problem; why don't you give the General a call?"

"I will once the contract is signed."

They received the contract documents just before 5 PM and spent all evening reading and marking various paragraphs. They found nothing offensive, simply a few nit-picks. Jon scanned the pages, including their annotations and emailed to Katie. He forwarded the originals because they contained the signature pages, and should the team be in agreement, UPS would be open on Saturday with overnight service. He then called Katie to alert her but had to leave a message. He wondered what they were up to.

Jon then emailed the scanned pages back to Margaret and the General.

By mid-afternoon on Monday it was all done. Jon had delivered the original signature pages to the Pentagon and had a brief meeting with the control group's new leader, one-star Gus.

He listened to Jon's requested use of the com link and placement of a special antenna at the base. He gave his OK and said Gloria would be in touch to coordinate everything including transportation. Jon said he would be calling Don and Charlie later that night.

They shook hands and Gus said, "Let's go get 'em."

Upon Jon's return to their office, Michael summoned both him and Karl. He had been utilizing their VPN, now with Washington DC as a node, to review the database in Cabo.

Jon and Karl read the call transcript. Karl asked, "What do you suppose that's about?"

Michael said, "No idea but I don't like it. Can we use this system to find out where the meeting will be?"

Jon answered, "Let me think about that. In the meantime determine the spot beam settings for all of Sinaloa Province and identify the cell towers. Guzman won't be leaving there anytime soon."

Randy was fully settled in his new office trying to read and understand both the system architecture and code Jon had written. He had never before seen anything like it. The descriptive terms that came to his mind were complex, elegant, simplicity, a bit of an oxymoron he thought. Karl completed bringing up the servers and workstations, testing they were all connected to both the local area network and the VPN between their sites. The first of the three large rooms opposite the offices was fully functional. It contained everything necessary for control and operation of the first satellite that they were choosing to call GEOWest. The second room was a near replica intended for GEOEast once the equipment to be installed in Afghanistan was up and Gloria had set up the link. The third room was primarily for storage of the anticipated mass of intelligence. It also housed two printers and the control center for the security system Michael had installed.

The four of them were sitting in the conference room sharing a pizza for lunch. Randy asked, "You've furnished eight offices. What do you have in mind regarding personnel?"

Jon swallowed a large bite before answering, "One is yours obviously, and the two at the end are Karl's and mine. Should Don and Charlie join us, two more. The use of the other three is contingent on the amount of action we see, for lack of a better term."

"That makes sense and the reason I asked is because several people at NSA told me, as I was leaving, if I ever needed additional personnel, please consider them. There are a few dynamite, disgruntled people at NSA that I would highly recommend as analysts. Keep it in mind please."

"Will do and thanks. Always share your thoughts, concerns or whatever with us. That's how we operate."

Karl changed the subject, "I have all the coordinates you requested and after lunch thought I'd show Michael how to control GEOWest and see what we get. Join us Randy. Also it occurred to me from the transcribed calls we have, there's some intermediary Guzman uses in Mexico City. We have that phone number. Why don't we see just how good this system is by focusing the other antenna there and use that number as the filter? These servers need to earn their keep."

The evening phone calls between Jon and Katie were endearing, humorous, and frustrating all at the same time. They both missed each other having rarely been separated the past ten months. Katie admitted to the possibility of suffering from separation anxiety then asked, "Talk dirty to me."

Jon told her it should be no more than a week or ten days before they would be fully functional and he was tired of 'hurry up and wait'. When the new headquarters facility was ready he suggested an open house for the control group. She thought it a great idea and then added, "It's too bad Hector can't cater."

A day later, with some minor frequency tuning, they had intercepted seven calls from the Mexico City contact or relay as they were beginning to realize. There had been no calls of any significance, either cell or sat, from the Sinaloa antenna spot beam or the Cabo sat capture system. Deletion of the extraneous calls was currently requiring human intervention and Jon was sketching out possible solutions to automate the process when Gloria called.

"Hello, this is Gloria. Do you remember me? General Shepherd provided your number."

"Hi and yes I remember; General who?"

"Gus Shepherd from DIA."

Embarrassed, Jon said, "Forgive me I didn't remember his last name, if I ever knew it. Probably should have read the tag on his office door the other day."

"I understand we have some equipment to ship and install plus a pipe into the com link to the base. Is that right?"

"It is and let me find Karl Oberling and pass the phone. Give Randy a call regarding the link. He's part of us now if you didn't know. Thanks for the call and your help, Gloria."

They were sitting in the conference room discussing possible plans and logistics after Karl's and Randy's calls with Gloria. Jon turned towards Ricardo and asked, "How would you like to do us a big favor?"

"I want to help and be involved. What can I do?"

"Go back to the base in Afghanistan for a few days and help us install a new antenna and some other equipment."

"Say when," Ricardo answered.

"The day after tomorrow," Karl stated, "I changed shipping instructions from here to Andrews after Gloria's call. The equipment will be there tomorrow and speaking of tomorrow, Ricardo, we'll walk you through everything, although I know you've already observed most of it."

Michael interjected, "I don't know what else I can do to help and I'm ready to go home; thought I'd catch a commercial flight in the morning."

"I'd love to join you," Jon added.

Karl had taken Ricardo to the roof and walked him through the dish installation; the cable runs to their office and the additional connections to the backend computers. They discussed various scenarios for installation at the base and the needed modifications to the computer software and systems already there.

Ricardo asked, "Does Col. Davis know what I'm doing, and what about Don and Charlie?"

"Thanks for the wake-up call. I'll take care of briefing them."

In parallel Randy was working with Gloria connecting their office to the link from the DIA and to the Afghan base. Jon was evaluating various network latency solutions and waiting for that link to be connected for testing. Michael was on a flight to Tucson.

The rest of the day drug on and Jon was looking forward to his evening call with Katie. He decided to kill some time and went to the GEOWest room to see what, if anything, had been intercepted.

He scanned the logs then dug deeper into the transcripts. With a sudden sense of alarm or foreboding he accessed the Cabo database for comparison purposes. It was all there; he could sense it and almost see it. He yelled for Karl.

Randy heard the yell and arrived first; Karl then Ricardo followed soon thereafter.

"I've sorted these calls in order. Come over here and read the transcripts."

After everyone had a chance to read, Karl said, "We have a problem. Has Michael landed yet?"

The first call on the sorted list was to Guzman's relay from Zambada, who was identified by his known SAT phone number. It set the day and time of his meeting with El Chapo for eight days hence and asked about the meeting topic. The answer given was 'Reynolds and his friends'.

The second call was also from the relay to Guzman, just a sentence: meeting confirmed; date and time faxed to Jesus for delivery to you.

The third call was from the Zambada SAT phone to a new cell number with a request to secure lodging in Parral for three days on and either side of the meeting.

The fourth call was the fax from the relay to Jesus, no translation possible but recognizable by the audible tones. They now had his phone number and Jon made a mental note to consider how to intercept faxes.

The fifth call was the most alarming of all. It was between two cell phones and was an order from El Chapo to someone called Guerreros. The translation came back with some fighter issuing instructions to pull back, rest, re-arm and be prepared for new attacks in Los Janos and cross border. The system intercepted this call because all of Guzman's names and monikers were in the listening and filtering criteria.

The sixth and last call was from one of the numbers in call five to a new cell number. The message was 'orders issued Jefe, they will comply. We await your visit and further orders.' One cell tower for this call was in Parral; the other near Mazatlán. They now had Guzman's cell phone number.

Ricardo altered the translation of Guerreros slightly, explaining it was actually a warrior, and had been adopted as an honorary term by the small armies the various cartel bosses had assembled to protect and extend their interests. He was equally alarmed as Los Janos was close to Sr. Valenzuela's property.

Randy said, "I'm really sorry for the terrible news but your system is beyond belief, otherwise you would never see them coming."

"What are we going to do Jon?" Ricardo asked.

"You, my young friend, are going to Afghanistan; we need you there. I'm going to Mexico and kill every god damn one of those assholes and put an end to this. That's the first battle. Then I'm coming back here, turning this system on and beginning a new war. Enough is enough."

Karl said, "To quote Katie, you are one scary dude, but I'm with you. I'll get Michael on the phone and brief him. We need to get back to Tucson."

Jon's and Karl's charter to Tucson was waiting at Andrews. Ricardo and the equipment were aboard the C17 about to taxi and head east. Randy had driven them to Andrews and assured them he could handle the DC end of things.

They boarded the Lear; the clamshell door was closed, and four hours later touched down in Tucson where Katie and Marie were waiting without the usual welcoming smiles.

Part 7 – The War Begins

Chapter 38

Michael wasted no time after the phone call. He contacted Rueben Valenzuela first and with no apologies explained what was happening and described a proposed plan. Several refinements later, it was adopted. Next he called in a career long accumulation of favors, deliberately leaving out any law enforcement entity in Mexico. Satisfied with the efforts so far, he knew Jon would need convincing and he expected some additions or changes. He waited for their arrival. Corinna came to sit on Michael's lap and held him tight to her breast.

There was very little conversation on the drive from the airport to the condo units. Upon their arrival and after perfunctory greetings, Jon announced he wanted a drink and some food. Katie mixed a pitcher of margaritas and microwaved mini pizzas. While they ate, Michael presented his plan and the players he had recruited. To his utter surprise, Jon agreed to everything. His only addition was that Randy should be able to monitor the specific Guzman, Zambada meeting area and might add to their knowledge and preparations.

Karl made several interesting observations, "Shouldn't we assume their attacks would be in parallel? If we're at the Valenzuela estate, who'll be protecting this well known address? Also shouldn't we assume they have people already in the Tucson area?"

Jon answered before Michael, "As to these addresses, they will be vacant and we will hire and put into place more armed security personnel than anyone would likely challenge plus nearly constant police surveillance. The cameras and security systems Michael had installed here should be able to identify any potential attackers and I think we should review recent images to see if anyone has been scoping this location.

"As to an attack coming from across the border, there are only two possibilities: the attackers cross legally or illegally. I intend to convene the control group to brief them and enlist additional resources. Sheila, from our control group as she's part of Homeland Security, will be requested to implement a high level of scrutiny for the border crossings near here plus alert the border patrol. All these attackers, warriors or whatever the hell they are called, follow directions. Take out their directors and that's it. Does that address your concerns, Karl?"

"Yes it does."

Katie asked, "If this place is to be vacated and if the Valenzuela estate is to be the battleground, I don't understand where everyone will be? Understand, I will not agree to any separation whatsoever."

Michael looked to Jon for an answer.

Knowing either an argument or confrontation would be coming, Jon thought fast before responding.

"I see no reason why Sr. Valenzuela could not house this group. We should ask him now. And we're sure not driving there, especially not with what I'll be bringing. Right now I need some sleep."

The next morning Jon called Randy and requested he courier to General Gus the Guzman call transcripts they had captured and request a call with the control group ASAP. *We're now down to five days*, he thought.

Four and a half hours later the control group was assembled, a few including Jon, calling in from remote locations.

"Your meeting, Jon," the General said.

"By now I hope each of you have read and understand the implication of the call transcripts Randy sent over. Our system is essentially operational and we have all the relevant phone numbers, both sat and cell. Randy will be constantly monitoring and will copy you with everything.

"Our group considers this upcoming attack to be the first battle in a new war. We have no intention to take defensive positions at our properties here in Tucson or at the Valenzuela estate. We plan to take the battle to the enemy. Our intel abilities provide an enormous advantage to see them coming and you all know we are fighters but we're not the military. The many contacts Michael Ortiz has on both sides of the border have been invoked, but at present our plan, that I'm willing to describe, lacks needed resources.

"Please consider what, if any, assistance you can offer that meets policy, laws and international agreements. And should you be able, AG Gates and Mr. Prescott can't stay on the sidelines. I'll answer any questions you may have. Thanks for your prompt response."

With almost no hesitation Gus asked, "What do you need?"

"First and foremost, Shelia's support in providing a high level of border scrutiny as some number of Guzman's soldiers will be headed here. We're assuming some already are in Tucson. Further, Shelia must remain in constant contact with the teams Sgt. Ortiz will have in place here.

"Next, a helicopter ride from here to the Valenzuela estate for our group that has the ability to clear both US and Mexican airspace without incident. I can't ask for what would tip the scales – firepower."

Shelia spoke up, "I am more than willing and able to support your first request. How I go about it needs to be discussed within this group, for obvious reasons. Have you involved the Mexican authorities?"

"No, and also for obvious reasons."

Gus asked, "What is the timing you are anticipating? We can see the Guzman, Zambada meeting is in five days."

"We know from our constant monitoring these past five months, Guzman and Zambada assembled a small army that has been fighting Los Zetas, most recently near Chihuahua. We believe they will be the attackers. Any satellite images from that area and around the town of Parral showing unusual movements could be very helpful. Our best guess is the attack will be concurrent with the meeting."

"Please give me twenty-four hours to see what we may be able to do. God knows you deserve our support."

Jon disconnected the call and sat back. To all the questioning looks around him, he responded, "We'll see – twenty-four hours max, Gus said."

Michael, who been reviewing images from the various security cameras around the condo units, announced he believed they had in fact been cased and most recently yesterday. Katie's printer came to life and six pages containing camera images were soon on the dining table.

"Same two vehicles and they are also in several additional images from last week. See the clear license plate numbers? I'm going to drive over and have a heads-up conversation with the Captain and then the security company to plan their surveillance as soon as we hear from your Shelia person."

Randy called the next morning to inform that General Gus wanted another call in two hours. Apprehension could be seen on the group's faces, including Jon who glanced several times at his watch.

The call commenced at 11 AM Tucson time.

"Good morning, Jon," it was AG Gates' voice.

Gus spoke next, "Both Preston and Clint have joined us for this meeting. I know you like the bottom line, but you should know that Justice burned the midnight oil to determine what we might be able to do and the President was briefed because of our Mexican relationships.

"First, regarding Homeland Security and the Border Patrol, full alerts and rules of engagement have gone out this morning for a line 32 miles east of Nogales to 46 miles west. That covers all of the likely legal and illegal crossing points. Shelia only requires contact information with Sgt. Ortiz.

"Next, and I won't take time to describe some of our crazy laws like the Posse Comitatus Act of 1878 that was amended last year, the US Military cannot conduct a unilateral operation in Mexico. Unilateral meaning any interdiction without specific authorization and/or cooperation from Mexico – as we did last year against Guzman from La Paz.

"The National Guard is not so restricted given certain provocations. I'll get to those in a minute. Also they are cleared for Mexican airspace without a flight plan and do not need any landing permit on private property. Your ride has been arranged. We simply need GPS coordinates for where Sr. Valenzuela would like them to land.

"Back to the National Guard, who is technically under the control of a state's Governor, but report through the Pentagon. A short while ago the President contacted the Arizona Governor and requested National Guard support of the Border Patrol, vaguely alluding to a possible Cartel war spilling onto US soil. They reached an agreement.

"Now for the provocations I mentioned. There are several as it relates to your situation. First and most likely, although timing could be an issue: should Shelia's team be able to identify a border crossing of armed men and trace them to an attack on your Tucson residence, that would constitute authority to engage the source or sources. Second, they be fired upon or attacked in any way. Rest assured their choppers have more than adequate firepower, and last, with the specific permission of the Mexican government, probably supported by a request from Sr. Valenzuela.

" There you have it. It's the best we can do. The President is more than willing to contact his counterpart in Mexico City should that become part of your decision."

Jon absorbed it all in the context of their plan, "Speaking for our entire group, we really appreciate your support and Shelia, we are particularly in your debt. Randy will make the Michael Ortiz connection today and please allow me to brief our group including Sr. Valenzuela. I'll be back in touch tomorrow morning, day minus three. Thank you all."

That afternoon was a whirlwind of activity. Jon gave the entire assembled team a detailed report on the call with the control group. Everyone breathed an audible sigh of relief.

Katie asked, "How does this alter our plans?"

Jon responded, "Truth be told, sweetheart, our plans remain in limbo, but by tomorrow night they will be set and we will swing into action the next day – day minus two."

Michael packed up some papers including his notes, ready to drive to police headquarters for his meeting with the Captain. Randy had provided the contact with Shelia and she and Michael had a brief conversation. Everyone overheard she would be called back from the Captain's office.

"Save me a margarita for when I return, I'll probably need it," Michael waved bye as he closed the door.

Karl piped up, "Jon, how fast can we set up a Web site with voice over IP capability? I'm thinking Michael, the local cops, the private security team and Homeland Security could all use it to communicate between them and there wouldn't be any of the usual BS to wade through."

Lucinda, who had been quiet and listening, offered, "I just created a new Web site for one of my classes. It's now live. Maybe you could modify and use it. I can help."

Jon grinned widely, "Karl, you and your new colleague better get crackin'. Oh, don't forget to grab a nap. Remember your night will be spent with Ricardo on the other side of the planet, while I sleep soundly."

With a grin, Karl gave Jon the finger.

The other 'girls' now lounging on the deck, overheard all the conversations and just smiled at each other. Tensions were beginning to subside but underneath everyone knew what they faced.

Michael returned with an announcement, "Everything is in place." Lucinda's Website had been modified and up and running with Google as the host. Karl had purchased and installed a VoIP app. They beckoned Michael aside and showed him, suggesting he connect using his cell phone. He did and after a major thank-you, told them he would connect the other dots in the morning. As almost an afterthought, he asked if the security cameras at the condos could be connected. Lucinda responded she was on it.

Katie, Marie, and Corinna were busy in the kitchen. Stephanie was setting the table. Jon and Michael entered. Jon opened the refrigerator door.

"What are you doing?" Katie asked.

"Fixing Michael and us a pitcher of margaritas."

"Sorry guys, not tonight. We're having an all-American dinner with only a slight Mexican touch."

"And what's that?"

"BBQ'd hamburgers and French Fries with all the fixings. There's a rack of Pacifico in the refrigerator and limes are in the lower drawer. Enjoy, and sorry Michael. You could crank up the grill in about a half an hour, lover."

After dinner, the 'boys' and Lucinda retired to the condo used as a shared office and prepared for a long night. Ricardo would be standing by at 10 AM his time. Before midnight Tucson time, Lucinda had modified the Website to add a link button allowing real time viewing of the security cameras. Michael had tested the new button successfully and they both called it a night. Karl had been on several long calls with Ricardo, and at 3:40 AM Tucson time pronounced the new antenna and connected equipment was ready for testing when Randy came to work in DC.

Jon had been stretched out on the long sofa in the condo-office, drifting in and out. He opened his eyes and told Karl, "Great job, I'll take the morning shift with Randy, good night."

The next morning Jon awoke and fixed coffee quietly as everyone else was still asleep. He went to the office to call Randy. Together they brought up the DIA link to the Afghan base, logged into the new computers there, and controlled the antenna directed at the second satellite. Then they began control. Within an hour they were receiving signals, and after frequency range adjustments, compared the new signal capture with the existing base system. They were identical. They found it difficult to exchange high-fives over the phone. The DC based central system was 100% operational worldwide in the northern hemisphere.

Jon asked Randy to set up a three-way call with General Gus for noon DC time, and knowing that Don's and Charlie's tour would be up in two days, booked three first-class tickets from Kandahar to Kabul to Dubai to Washington DC, Dulles, for what he was calling day zero. First class for the first two legs was a pipe-dream, but at least better than military transport.

Katie was the next one to awaken, and seeing the brewed pot, poured herself a cup, and guessed where she would find Jon. It was what he called day minus two.

She found Jon and Michael in deep conversation in the office condo. They looked up and both smiled, each with their own type of admiration.

"May I join you?" she asked.

Jon arose, gave her a kiss and guided her to the sofa he had slept on.

Katie raised her coffee cup and said, "Cheers to day minus two. All of us are hoping we will have a solid plan, sooner rather than at the last minute."

Both Jon and Michael took the rebuke without any retort. They simply raised their coffee cups.

Jon looked at his watch and said, "I have a call with Gus in ten minutes that should settle all issues and allow the creation of a solid plan. You're welcome to stay."

Jon's use of the phrase 'solid plan' was not lost on Katie, and feeling somewhat bad about her remark, she got up and left. Michael followed. Jon called the conference number right on time and he and Randy were connected to General Gus.

Jon began, "Our group is ready and prepared for what might be coming. Here are our decisions and a few last minute support requests. We will need two choppers as the number of people to be transported is 23. Rather than us buying out the local gun stores today, can we buy, beg, borrow, or steal ordnance from you or the Guard?" A chuckle followed.

"Next, instead of having the President call his Mexican counterpart, Michael Ortiz has a relationship with a trusted local Mexican police commander. At the appropriate moment, he will be contacted to clean up any mess and take control and custody of those left alive. Sorry to be so blunt but I know you, of all people, understand.

"And last we request the National Guard transportation either stay with their choppers and not engage, or take off to be summoned to return for a pickup. That's all. Oh, one more item: Randy will deliver to you three first-class tickets for Don, Charlie, and Ricardo. Our new system is 100% operational."

Gus replied, "I understand, will convey your requests and decisions and reply through Randy with any follow-up. Shelia has already been in contact with Ortiz this morning and the border monitoring is underway. Regarding ordnance, have a conversation with Lt. Ralph Simonson when you arrive at the Guard headquarters. I wish you well and look forward to your teams' return to DC, intact."

Chapter 39

Jon emerged from the office condo and summoned Michael to the deck. He closed the deck sliding door and they became engaged in a long, private conversation. All the others had awakened and were sipping coffee or juice and munching on delights from a plate of pastries on the counter, intensely curious as to what was happening.

Michael, somewhat out of character, first opened the deck sliding door and announced, "We now have a complete plan and are ready to let all of you know the details. Please be seated and comfortable. I wish to avoid any arguments or discussions. The plan is not up for a vote."

Nobody took offense to the change in their usual decision making process. They complied and when seated Michael began. Jon remained leaning against the open deck door.

"The Guzman, Zambada meeting is scheduled for the day after tomorrow. US recon satellites have noted the creation of a new encampment two miles from Los Janos. That is seventeen miles from the Valenzuela estate. Best estimate is there are about 280 people in this camp. Our systems' cell phone capture has confirmed the set-up and existence of this camp.

"Sr. Valenzuela has graciously agreed to host this entire group. We leave tomorrow morning via National Guard helicopters. In addition he will be hosting, or housing is probably a better word, 17 others consisting of family members and trusted neighbors.

"Homeland Security and the Border Patrol are currently monitoring many miles of the border, either side of Nogales. All our properties and friends here are under 24 hour a day protective surveillance, both private and police. All the agencies are in communication and cooperating, thanks to Lucinda. Through the license plate numbers from our security system cameras, two local cartel supported gangs have been identified, and should they so much as jaywalk, they will cease to exist. We will bring this to closure, God rest my soul. Jon? Did I miss anything?"

Jon moved into the room and leaned against the counter, "Let me think on that; the floor is now open for discussion as the Congressional jerks like to say."

Always the one to break the silence or alter the mood, Katie asked in a demur tone of voice, "Should I pack sunscreen and a bikini? And where do I get my automatic weapon with lots of ammo?"

Jon lost it and nearly choked laughing. The others followed suit. It took almost ten minutes to return to seriousness.

Jon regained personal control and said, "A couple of items Michael didn't mention: our new system in DC is fully operational. Randy will be monitoring almost 24/7, and we have directed the GEOWest spot beams to cover all relevant cell towers here and in Sonora Province. The Cabo system will continue to monitor the SAT phone range. With that we should see them coming as orders must be issued. Oh, and Ricardo, Don, and Charlie will be returning the day after tomorrow."

Marie asked, "The men intending to attack us will certainly be heavily armed. How will we fight back? I don't see any weapons we have."

"We had originally planned to solve that problem this afternoon, but it won't be necessary because General Gus Shepherd is our new best friend. You will see tomorrow."

"What can I do?" Lucinda asked. Stephanie echoed the same question.

Karl said, "I was wondering the same thing. What are the plans, if any, for what the military calls deployment?"

Michael answered, "We thought it best to have that discussion and resulting plan with Sr. Valenzuela tomorrow. It is his estate that will be attacked. Let's be real frank here. This fight is for our lives and the risks are high. I am a trained cop. We all know Jon's skills. Beyond us only three of you have weapons experience. Three do not. Marie, Stephanie and Lucinda, no one is going to insist you hide out, but we must be realistic. Tomorrow afternoon each of you will receive a handgun and training, and I recommend you remain inside the main house and protect all possible entrances.

"In addition to the 17 Sr. Valenzuela will have, I've called in many favors and will be traveling with 15 professionals. I really like our odds, especially with the intel we will have."

The magnitude of what was coming slowly sunk in and heads nodded. The air of determination could almost be felt.

The next day they drove in two crowded vehicles to the nearby Davis–Monthan Air Force Base, were passed through the gate because they were expected, and pulled up alongside a large helicopter. Their duffle bags and other gear were loaded aboard with flight crew assistance. Another large helicopter was nearby with doors open. Across the tarmac could be seen two approaching trucks and a uniformed officer on foot. The trucks pulled up to the other helicopter, and the officer continued towards the group.

Michael left the group and walked over to the trucks and began to greet the men climbing out.

"My name is Ralph Simonson with the National Guard. Which one of you is Dr. Reynolds?"

Jon stepped forward and offered his hand. "I am, and why do you seem so familiar?"

Simonson's mouth twitched into a small smile, "That's because General Shepherd is my uncle. My mother was the General's twin. Walk over here and I'll show you the ordnance we have provided for your use. We would like as many as possible returned."

The Lt. raised the helicopter's rear cargo bay door, reached in and flipped a light switch. Three wooden boxes were side by side, about the size of coffins. The Lt. unlatched the closet one and raised its lid.

Jon stared in disbelief. Inside was a very scary collection of automatic weapons and neatly packed ammunition.

"I took the liberty to speak with Sgt. Ortiz concerning his training and knowledge. He described your skills. The adjacent box contains handguns and sniper rifles. The other box contains additional ammunition and several spotting scopes. I hope this meets with your approval."

Jon reached out his hand and said, "Beyond my wildest imagination. Thank you."

"You're welcome; better get on board."

After parking the vehicles, the entire team boarded and the whirly birds lifted off for a short flight. Jon whispered to Katie, *It's like back at the cabin. Are you ready sweetheart?*

Katie squeezed his hand and said out loud, "Damn straight."

Sr. Valenzuela heard the helicopters approach and stood in his large, groomed rear yard as they descended to their predetermined pads. Thirty-eight people began unloading gear onto two flatbed pickups that Valenzuela's sons had driven over. Once completed, they cleared the area as the helicopters wound up their blades. The choppers were off and out of sight within minutes, prepared to return on an hour's notice.

Valenzuela's sons were greeted warmly. It seemed Christmas was a very long time ago. One of the sons would take Michael's men to the bunkhouse where others were housed. Jon requested the three wooden boxes be left on the pickup and not opened.

"Miguel, allow me to show you your accommodations, and you are most welcome here. After unpacking please join father and I on the patio."

The upper level of the Valenzuela house was vast. Katie remembered from Christmas the number of offspring and other relatives, so the number of rooms should not have been a surprise. Each had their own bedroom with a lav and toilet. Once settled in they gathered outside as the temperature was less than ninety degrees.

Sr. Valenzuela came out from the house and greeted them with a warm smile, "Please be seated and some refreshments will arrive shortly."

Lucinda hugged her patron and mentor tightly for almost a minute.

All the hugs, greetings and introductions were finally completed. Karl, Marie and Stephanie had heard about the patron they now faced and his estate, but were uncertain if any of what they observed was real. It was another world for them.

"Welcome and please know I do not place blame. This is not only because of your actions that I do support; it has been building for many years."

Refreshments arrived and were served. They sat around chatting as if there wasn't a care in the world. Lucinda was seated next to her mentor. Jon and Katie were on the other side. Lucinda leaned over and asked Jon how much she could tell about their activities, remembering the trip to Skazar was off limits.

"Go for it, no restrictions, and don't leave your brother out."

She grinned and began engaging Sr. Valenzuela in conversation. He seemed very pleased.

Two hours later Michael disengaged from his and Corinna's conversation with one of the sons and announced, "OK folks, back to the real world. We have work to do."

Suspecting what was in the wooden boxes, he added, "Jon, let's see what you brought us."

Everyone gathered around the flatbed pickup and with a lot of effort, lowered the boxes to the ground.

Katie quipped, "You know what those look like."

Jon nodded and opened the lids on each.

"Oh my God!" exclaimed Marie. The rest were speechless.

Michael reached into the center box and extracted three Glock 19s and two boxes of 9 mm ammunition.

The slides were open as he handed one each to Marie, Stephanie and Lucinda, keeping the ammunition.

Addressing Sr. Valenzuela, he asked if there was a hay bale they could use and maybe some type of target.

The youngest son, Ramon, answered, "Please get in the truck and I'll show you."

The three women squeezed into the small rear seat with wild eyed looks. Michael was in the passenger seat. They stopped at a maintenance building and Ramon fetched several bulls-eye ringed paper targets. A quarter mile later, hay bales stacked ten feet high could be seen.

The crash course lasted almost two hours. Michael was in his element showing them everything prior to pulling a trigger. They loaded, unloaded, practiced positions, assembled, disassembled many times. He then demonstrated grip and various methods of aim, depending on a given situation. At last the targets were set and he scraped a firing line in the dirt with his boot, 100 feet away.

"Ladies, at the risk of offending you, please remember how you like your nipples squeezed. That is exactly the speed and pressure you will apply to the trigger. Stephanie, you are first. Please step up here. You know the standing position."

Stephanie, showing no apprehension, took her stance and aimed. Michael nodded and she squeezed the trigger. There was only a slight kick and a hole could be seen in the upper left part of the target.

Michael held out his arms and Stephanie ran into them with a huge grin. Both Lucinda and Marie did just as well. Over the last half hour, near constant gunshots could be heard. Aim and positions were adjusted. Lastly, they practiced while moving, and from the hip as if in an emergency situation.

Upon their arrival back at the Hacienda, they piled out of the truck and began talking at once to whoever would listen, obviously ecstatic.

"OK Katie, Corinna and Karl, time to practice your skills before the cocktail hour dulls your reflexes. What might be your weapon of choice? Jon, what do you think?"

"I trained Katie in both hand weapons and rifles. What's your choice Katie?"

"I want a Glock like them, and one of those machine guns if Michael can show me how to stay on my feet."

Karl said, "Same as Katie."

Corinna said, "A Glock and a sniper rifle. I'm actually a pretty good shot."

The latest trio piled into the truck with Michael and Ramon and off they went.

Soon the sounds of automatic firing were heard across the estate. All twenty-six of the men lounging around the bunk house were on alert. Jon ran over to where he could be seen and waved them over.

Ruben Valenzuela raised his hand and quiet ensued.

Jon asked Valenzuela if English would be OK and would he translate as needed. He received a go ahead nod.

"My name is Jon Reynolds. I am your source of information and direction, but not your commander. I am deeply appreciative of your willingness to support my group, which includes Michael Ortiz and Sr. Valenzuela in the protection of his magnificent estate and our lives."

Jon hesitated and Valenzuela translated.

"I will be totally honest with you and if any of what I'm about to disclose causes alarm or second thoughts, you are free to leave with our best wishes. We believe an attack is coming sometime after an El Chapo, El Mayo meeting tomorrow. The attackers will be from the small army those two formed to fight Los Zetas. The attack will come from Los Janos where they are currently camped. For reasons not relevant any engagement must be on Valenzuela estate soil – that is a requirement.

Jon hesitated and Valenzuela translated.

"Please step forward and select any additional armament you might desire, however we must have your name and the weapon taken, as we request its return if at all possible."

The men circled the boxes almost as much in awe as previous viewers. One man reached down and picked up an M-16 assault rifle and looked up at Jon, who nodded an OK. Within ten minutes the boxes were 80% empty. Valenzuela's oldest son was writing down names and weapons in a notebook.

"Our group or team has many intelligence sources. We don't anticipate any attack prior to the meeting, only possible movement and preparations. We will meet here at 10 AM for deployment instructions, please be ready and thank you again for your support.

Valenzuela translated and the men returned to the bunk house awaiting a promised dinner. Jon and Ruben Valenzuela sat alone in comfortable silence.

Jon could hear the approaching truck and asked if he could invade the bar and fix margaritas and Sangria, claiming a certain expertise. Valenzuela laughed out loud and gave his permission, warning that his housekeeper would be on Jon in a flash.

Katie, Corinna and Karl bailed out of the pickup just as excited as the prior trainees. Michael looked exhausted.

Katie looked around, "Where's Jon? I need to tell him how much lead I put in the bad guys."

Valenzuela replied, snickering, "Fixing some American version of our national drinks."

Just then Jon and the housekeeper emerged with pitchers, glasses and a large tray of snacks. They proceeded to serve everyone. When a glass of Sangria was handed to Sr. Valenzuela, as he did not drink hard alcohol, Jon smiled a 'Please'. The housekeeper was keeping a close eye out as he took a sip.

"What's that American expression? Old Dogs can learn New Tricks? This is very good; I'm impressed."

The housekeeper looked disgruntled.

Stephanie and Lucinda were also given Jon's Sangria and felt like they had finally been accepted as equals into the group's world.

Jon had taken Michael aside and described the communication and weapons exchange with their 'army' and the meeting for the following morning. He also reported Randy had not intercepted anything of importance and was going home for dinner and a few hours' sleep before day minus zero.

Chapter 40

It was a restless night for all, as they went to bed and their brains would not let go of what the new day would hold. The evening had been pleasant, with everyone except Jon and Michael, buzzed on the drinks and sated from a delicious dinner, enjoying the warm temperature and relating their weapons training and practice.

Jon, Michael, Valenzuela, and both sons huddled over maps of the estate, including topography and the various routes from Los Janos.

Valenzuela said, "Thinking as I believe Guzman and Zambada do, and as a contingency to any intelligence you obtain, let me describe how I would attack this property."

He picked up a marking pen and drew several lines, followed by several circles.

"The lines represent incoming routes. The circles represent our defensive positions. The only uncertain area is along the creek bed behind this building, if it is even known."

Michael looked at the annotated map for several minutes and responded, "I trust and defer to your knowledge. That is where we will deploy. However we organize the men in the morning, there must be a leader at each location with a radio communicator and sufficient vehicles to move quickly."

"Agreed, it will be done."

Jon was awakened to the beep of his phone announcing an incoming email. He quietly rolled out of bed and read the message. It was from Randy.

Randy answered on the second ring and announced, "We have movement."

"Please summarize."

"Zambada is already in Parral, Guzman is expected by noon. Their army, for lack of a better term, camped near Los Janos, received instructions to close up the camp and stand by for additional instructions.

"Border Patrol reported observing a truck unloading eight heavily armed men on US soil, at a crossing twelve miles west of Tucson. Shelia instructed not to engage and they were picked up by three SUV's, one with the license plate you previously provided. Tracking them is in progress."

"Randy, keep up your outstanding support, many thanks, and we're ready to engage here."

At 10 AM sharp everyone was assembled around the Hacienda patio area, weapons in hand and ready for direction.

Michael stepped out in front, "We need to break up into six teams of equal size. Please take a moment to do so and identify your team leader or communicator."

It was done in less than five minutes with the identified leader in front. Jon walked around and handed out two-way walkie-talkies, fully charged and preset to channel 7.

Michael continued, "These radios must be monitored all the time, and we ask you to comply with any directions immediately, for your own safety as well as ours. Only we have radios, so you will know it's us. We have received intel that the camp at Los Janos is being closed up in preparation for a move. Jon will now hand out your deployment locations. Please set up there ASAP, be observant, and await radioed instructions. Many thanks, guys."

All the chosen team leaders spoke English so no translation was required. Sr. Valenzuela had supplemented their personal vehicles and after several minutes of discussion departed. Michael's men were disbursed among the teams as they lacked local geographical knowledge.

Turning towards Sr. Valenzuela, Michael said, "With no disrespect meant, you, Marie, Stephanie, and Lucinda will protect the Hacienda. I leave it up to you to deploy as you see fit, and Marie's cell phone, which is part of our network, will be the only method of communication. Your comment about the creek from last night was not lost on me.

"Are we ready to retake our lives? I sure as hell am!"

Katie walked over to Marie, Stephanie, and Lucinda and engaged them in a group hug, whispering, *'Be strong and tough. You can do it.'*

She beckoned Corinna, then grabbed Karl's hand and with her Glock in her waistband and an M-16 over her shoulder, walked up to face Jon and Michael. She was almost as scary as Jon was reputed to be – some kind of '*Wonder Woman*'.

"Boy's, it's beyond time to kick some serious ass. I'm not living this way anymore."

Guzman arrived in Parral just before 1 PM local time and was let out in front of the Café los Corrales, the site of the meeting. Zambada was already inside and seated. Upon Guzman's entrance he stood up and greeted him warmly.

"Please sit and enjoy the food – it's alright. As you requested, our forces are ready. Explain what we need to do and why."

Guzman sat and drained a glass of water before he spoke, "Do you know why your ship returned from Pakistan empty? Do you know why the Afghan sources have dried up? Do you know why I'm moving my facilities so often? Do you know why every one of my residences and methods of communication changed? Are you so naïve as to believe you're safe in Panama? Do you know why I lost over $400 million in the past year and you didn't lose a penny?"

El Mayo was rapidly thinking about how to answer those questions, when El Chapo continued.

"Allow me to provide the answers. First, two people and their associates: Jonathon Reynolds and Katie Ruiz, media darlings. Second, you made a deal through Ruben Valenzuela. Don't deny as I have confirmed information. Those are the facts.

"Do I blame or resent you? Somewhat, but given the same sources and opportunity, I probably would have done the same. We both know demand for our products continues unabated in both North America and Europe. Thanks to our mutual efforts, Los Zetas and any other serious competition is dead – one of my favorite words.

"That brings me to this meeting and the road forward. Our recent partnership has been mutually beneficial, excepting three things: sources of supply, intercepted communications, and lack of candor. How would you like to continue, partner in name?"

Zambada decided placation would be best, "It appears you already have the answer to that question. I and my organization have become dependent on you and I'm comfortable with the situation, as long as we are in harmony. What must we do?"

"I want Valenzuela and every other unsupportive land owner destroyed. I want Reynolds, Ruiz and their entire group of associates eliminated. I want everyone, including the remnants of Los Zetas and the other cartels to know that, 'Nobody fucks with El Chapo'. How can I be clearer?"

Zambada already knew how revengeful Guzman could be, but instantly decided it was in his best interest to go along.

"Let's give the orders and have a drink."

Jon answered Randy's call on the second ring.

"Jon, it's going down. Do you want to take notes?"

"No, please just tell me."

"Two cell calls just went out, both from the town of Parral, both from Guzman's cell number. The first said to attack the Valenzuela estate now. The second told an unknown recipient to attack the Ruiz property now. That last call was linked to a cell tower inside the Tucson city limits. I have alerted Shelia already."

Jon explained the call contents to the group and said, "The battle begins; let's get in position. Michael, alert the other teams by radio and call your Nogales contact now."

He nodded as they climbed into another Valenzuela truck. Michael was in the back seat on the radio and then the phone. With Jon driving they were at the southern edge of the Valenzuela estate within ten minutes.

Michael had a long standing relationship with the Chief of the Nogales police and had alerted him to be prepared for a call on this day. He answered on the third ring.

"Greetings and thanks for taking this call. How you choose to react and respond to the information I'm about to give is entirely up to you. There will be no other information provided."

Michael had the Chief's attention.

"If you want to capture Guzman and Zambada, they are sitting in a restaurant in Parral at this moment. There will be a battle beginning shortly, along the southern edge of the Valenzuela property. Stay away and tomorrow please clean up the bodies. That request is my price for the previous information. Sorry, can't stay on the line. Best regards."

Jon, Katie, Karl and Corinna were in a prone position on a knoll, amongst some trees, overlooking the main gravel road entering the Valenzuela estate. Dust in the distance signaled the arrival of someone. Soon three trucks were visible. The first was a flatbed pickup with 50 mm gun mounted behind the cab, manned by a gunner. The following two trucks were larger and the space behind the cab was canvas covered. They were traveling at about 20 miles per hour.

Jon looked at Corinna and she smiled back. Almost simultaneously, Jon took out the driver and Corinna the gunner with single shots from their sniper rifles. The trucks came to an abrupt halt and many armed men could be seen jumping out of the following trucks and dispersing on either side of the road, maybe forty in all.

Michael came up from behind them and signaled Karl to follow him. They disappeared over the knoll to the right. Jon turned to Katie and indicated they would do the same to the left.

He looked down at Corinna, still prone, and asked, "Are you OK?"

Just as she responded, "Get going," weapons fire could be heard in the distance.

Jon and Katie went left and crossed the road keeping a watchful eye. They entered another small group of trees and hunkered down. They could hear voices shouting close by. The minutes passed and the voices became louder, plus the sound of feet on the ground.

"Get ready, love," Jon said quietly.

"I am."

Several minutes later, Jon whispered, "Do you see them?"

"Yes, I just counted eighteen that are spread out. What should we do?"

Jon was about to respond when the sound of automatic fire could be heard from the other side of the road. Michael and Karl, Jon wondered? That sound caused the approaching men to stop. Jon and Katie waited.

Jon leaned close and said in her ear, "See that tall cottonwood tree on our left?" She nodded. "When they get up to that point, rise to your knees, do not stand, pull the trigger and spray from their center to your right. I'll do the same left."

The men began moving forward again.

Jon watched and said, "Now."

It was no contest. Eighteen dead bodies never knew what hit them. They remained silent, training all their senses to the area around them. After almost ten minutes they returned to Corinna's location and found Michael and Karl already there.

With a barely suppressed gloat, Karl stated, "Our body count is 23. How did you guys do?"

Before they could indulge in self-congratulations, four more loaded canvas covered trucks could be seen coming their way. Muttering 'enough is enough', Jon fetched a canvas bag and a large plastic box from their hidden truck.

Opening the bag and looking at what Jon had extracted, Katie exclaimed, "What in the hell is that?"

"An M-72 Laws rocket. Help me assemble the tripod, quick and get the other shells ready."

It was set up and loaded before the first truck was in its short range. In a prone position, Jon took aim and pulled the trigger. He felt no sense of elation or any remorse when the lead truck exploded. Quickly, as if already knowing what to do, Michael provided the next shell, and the next, and the next. The trucks and whoever was in them ceased to exist.

Chapter 41

From deployment to conclusion, the battle had lasted two hours and forty minutes. The total enemy body count was uncertain, but best guess from all the intact, returned teams was close to 300. Most of the group was slumped in chairs on the Hacienda patio. The ones left to guard the Valenzuela home were disappointed in not having any action, but were happy to have everyone back safe.

Jon had two missed calls; Michael had one. So as not to talk over the top of each other, Jon said, "You first."

Michael's missed call was from the Nogales Police Chief.

"Miguel, we are in your debt again. Both El Chapo and El Mayo are in custody. For how long is not my problem, thank goodness. There were vague rumors around Parral about a cartel battle. I will investigate tomorrow, and as you requested, cleanup. I'm not sure I can deal with my new celebrity status, but thanks anyway for that. Goodbye for now."

Michael just laughed, hung up, and repeated the conversation for the group.

Jon returned Randy's missed call first. He wanted to pass along a message from Shelia, the second missed call. "There was a confrontation at your Tucson property. It's over. You should give her a call."

He did, and Shelia repeated Randy's circumspect words and suggested the best person to call was the Tucson Police Captain.

Jon hung up, and repeated the conversation for the group. He asked Michael to make the call and put it on speaker so everyone could hear. Michael did.

The Captain came on the line and requested he put them on hold while he located Cody. Then he came back and said he would call them back in a couple of minutes. They waited for over ten minutes before the phone rang.

"Cody is now here with me as he conducted the operation. Yesterday we received a communication through your amazing Website that identified eight armed Mexican nationals being picked up by two local gangs. Your Website also allowed us to identify them, again amazing. With the assistance of Border Patrol we followed them to a local house. By the way, it has been raided and shut down. Over a million dollars in drugs were found, and the two connected gangs are now in custody, and won't see the light of day anytime soon.

"Cody was on surveillance at your property and in constant contact with your private security people and that Homeland Security woman. Through your private security people and HS we were alerted that a certain rented vehicle had passed slowly by your property several times.

"Cody and four other officers took charge. At the time of their arrival, the perps were attempting to pry open the side main door. When confronted, they pulled weapons but were shot in place. Cody, do you want to add anything?"

"Hi Sarge, we're just glad we could help. The Captain summarized it accurately. I want to echo his amazement about your Website and the communications methods it provided. Maybe there's a part-time retirement career in your future."

"Thanks for the protection and information you provided. We all really thank you. About the Website Cody, I'll introduce you to the woman who created it."

"I can hardly wait Sarge, bye."

Almost in unison the 'girls' began to clap their hands and smile broadly. Lucinda, while clapping, had a look on her face that read: *what just happened?*

Jon was nonplussed and focused his attention on their host, "What can I do to thank you, your family and friends, for their support?"

Ruben Valenzuela simply waved his hand in a brush off manner that said, 'nothing'.

Jon was not about to take no for an answer, "Would you accept a ten million dollar endowment for the education and betterment of your people?"

Valenzuela appeared to take Jon seriously, "Would there be conditions?"

"None whatsoever."

"Then I will accept. I know you need to do this as much as my people need to receive the benefits of your gift -muchas gracias, amigo."

Jon reached out his hand. Sr. Valenzuela grasped it firmly. The applause resumed.

All the weapons were returned and packed in the coffin like boxes. Most of the ammunition was gone. The helicopters would be arriving in less than an hour. The men in the bunkhouse could be seen exchanging good-byes and promises to keep in touch. Jon had given Michael an open checkbook to compensate the team he had recruited as he saw fit.

Sr. Valenzuela was standing near the pile of gear as they all heard the helicopters' approach. His surrogate son, Michael, Miguel, gave him a loving embrace. So did Katie, Marie, Stephanie, Corinna, and Lucinda. He couldn't remember the last time he had received such feminine attention. The process was repeated for his sons. They too were taken aback.

Jon simply shook his hand and said, "I will always have your back as you have mine. See you sooner rather than later."

The gear was loaded and the helicopters returned them to Tucson. Upon their arrival, Lt. Simonson was on the tarmac to greet them. Jon was pleased to report all weapons had been returned but he wasn't able to pick up the brass. The Lt. got a large chuckle out of that.

He said, "I don't suppose you've seen the news. Probably exaggerated with all the usual speculation, but from what I know congratulations on a terrific hunting expedition."

There were a large number of bloodstains on the concrete walkway beside the garage to the front door. As it was aggregate, Katie restrained herself from being pissed about what the cleaning would entail. She, along with the others, was happy to be home, safe and alive. Because of the hour pizza was ordered in.

The food arrived and consumed without much comment. It seemed everyone was being extremely introspective and soon said a perfunctory goodnight, including Jon. Katie was left alone. She fixed herself a nightcap and went out to the deck and the city views below. Her non-stop brain refused to shut down for the night. Memories came to the forefront:

Mrs. Reynolds I am, she thought, truly amazing. The first time she saw Jon, he was fumbling to find a mouse in front of an ancient terminal in the Arizona Star archives. Both their first and honeymoon hikes in the Saguaro National Park. The first battle they had against the drug cartels at Jon's Idaho cabin and all her fears. The TV interviews, dodging the media and the Medal of Freedom ceremony, and then the faces of all her friends passed by her mind's eye. What a ride.

Katie finally decided to call it a night and set up the morning coffee, leaving a sticky note that read 'just push the button.' She crawled in bed carefully so as not to disturb Jon and fell fast asleep, at peace with herself.

Katie was the last to make an appearance the next morning. She went to pour a cup of coffee and found there was none. Looking around at the dull faces, she couldn't resist saying.

"Gee thanks guys and in particular to whoever emptied this pot. Don't suppose it ever occurred to you someone beyond me, like a salesperson or reporter, might knock on the door and want a cup of coffee."

They fell all over themselves to make amends, but Katie simply went to the refrigerator and poured herself a glass of OJ, effectively blowing all of them off. Jon went to the coffee maker to create a new pot.

"Don't bother, too little, too late, unless of course anyone else needs more caffeine to wake up. Good morning."

She refilled her OJ and said to no one in particular, "Guess it's time to turn on my phone and laptop and see what the outside world has to say."

Jon said softly, "You're not going to like what you see."

"I'll second that," Michael added.

"Then maybe I won't look and become a recluse."

"Fat chance," Karl said from across the room, "we're going to take our leave and be in our unit if needed. On second thought, with all the decisions we need to make, maybe the recluse idea is not too bad."

Katie sat at her desk and booted her phone and computer. Then she began to search and read. The phone contained five voice messages. Her email inbox had seven more, with most a duplicate of the phone calls.

Jon went to the shower. Everyone else cleared out leaving Katie alone.

The lead story in nearly every media outlet was the capture of Guzman and Zambada along with three unidentified others. They had been turned over to the Federal Police and were being held in a facility near Chihuahua. The Nogales Chief was quoted to say he acted upon a reliable tip, refusing to identify the source. The reports tried to connect the human carnage, found north of Los Janos, to the arrests. There were many sensationalized statements like a cartel war never before seen, and over 300 dead, bullet ravaged, bodies strewn across several square miles. The Nogales Chief had announced they were at the site investigating. The media kept referring to it as a 'war' and the published pictures did look like a war zone.

Continuing to read, Katie did not find any speculation. The big unknown for the media was, if the bodies were the losers in that war, who or what cartel was the winner and what was the motive. And then she read a sentence from one report questioning whether it was in fact a cartel war.

There was no way to bury Cody's actions as many 911 calls had come from Katie's neighbors. The Tucson police department had issued a formal statement describing the incident and identifying the dead as Mexican illegals believed to be associated with one of the cartels. The location given was the '200 Block of E Guthrie Mountain Pl'. Katie went on instant alert.

An Associated Press reporter was the first to recognize the location as being near or at Kate Ruiz's residence. In a story released about the time Katie was getting her OJ, the speculation hit the fan. Not labeled as an opinion piece, the report stated the location of the Tucson police's action was at the residence of *Arizona Star* reporter Kate Ruiz. It went on, attempting to connect the dots by assuming the timing of all the events was more than just a coincidence, and the dead bodies in Mexico were of the same ilk and source as those in Tucson. The *Arizona Star* was asked to comment.

Jon came out of the bedroom and waited for Katie to say something.

She turned in her chair to face him, "Should we have anticipated this?"

"Yes and no. My only concern is a leak from any of those directly involved; the Valenzuela family? I think not. Michael's men? It's possible for sure, and he should get on it immediately. Knowledge is very compartmentalized, and outside of that nothing directly connects to us. We simply have nothing to say and ride it out, but should expect more speculation. Let's all get together and reach a consensus."

The first phone message Katie returned was Edgar Greeley. She put his fears to rest and suggested he not respond to anything, claiming no knowledge. The second was Ricardo, now in DC, who said he and Jon had spoken earlier and all is well. The last two she decided to respond with a generic email: AG Gates and Barbara Walters.

It was almost noon and Katie had just finished getting dressed when her front door buzzer rang. Peering through the security hole she saw two men, one with a camera bag, obviously reporters. She opened the door leaving the chain in place, "How can I help you?"

The shorter man, without the camera bag, said, "We are with Arizona Foothills magazine, and thank goodness you are all right." Katie thought she would gag.

"Our readers would love to hear from you, the area's most famous person. Please grant us an interview. You may edit any content before publication." Yeah, right, thought Katie, "Sorry but no. Thanks for your well wishes. Have a good day." She gently closed the door.

"Who was that?" asked Jon. She told him.

"I'm hungry; what's to eat around here?"

"I was just over at Marie's and Karl's unit. She has a brunch prepared. Let's go. Everyone is there."

They ate a near perfect replica of their breakfasts at Cabo and then got down to business and needed decision making.

Jon opened with, "I have related our earlier conversation and my conclusions. Everyone has seen all the media reports and Michael will be contacting his men and Sr. Valenzuela this afternoon. That's about as tight as we can make it. General Gus and the control group want a briefing, but that will be in person.

"Katie previously suggested an open house at the DC facility. I think that's a great idea, so what else?"

Corinna said, "Our Alaska cruise is coming up. Michael and I plan to lay low here until then, unless we're needed for something else."

Jon looked around at various head shakes and responded, "No, just enjoy your new retirement and trip. We will always keep you in the loop."

Marie spoke up, "I'm really looking forward to being back home as a housewife again, completing some details on our remodel. You really don't need me at the open house. I know you and Karl have to get the new system and business up and running. I'll just lay low like Corinna and see what shakes out this summer."

"I want to stay involved, if possible," Lucinda said.

"So when do we leave for Washington DC?" Katie asked, "I'm ready to fulfill all my Managing Member duties, ha, ha."

Chapter 42

It was summer and it was hot. It felt hotter than Tucson, but that was because of the DC humidity hovering above 90%. They had flown into Andrews on a Lear charter and been taken directly to their office to prevent any possible recognition. Media speculation continued, claiming possible connections of the Mexican and Tucson events with both Jon and Katie. So far Karl and Michael, known associates, were off the radar. The most troubling story was in the Huffington Post and began with the headline:

Was it Revenge?
If it was, from What Direction?

The troubling aspect was the underlying story was very close to accurate, even though it lacked any supporting facts. The byline was the woman who had given Katie her card at the restaurant.

Jon, Katie, Karl and Lucinda were joined in the large conference room by Randy and Ricardo, plus Don and Charlie. They were discussing employment details including roles and responsibilities. The DIA and General Gus had signed off on Don and Charlie moving to active reserve status. Katie was reading a memo from Randy about office staffing along with his recommendations. Katie put down the memo and Lucinda reached over to pick it up, "May I?" Katie nodded OK.

Karl and Randy took Don and Charlie on a detailed show-and-tell tour of the facility and the systems it housed. Jon went online and found a DC area Cadillac dealership. He picked up his cell that did not broadcast Caller ID, called the number and asked for sales.

"I recently saw the CTS advertised and am interested. Do you have a dark blue color and what is the MSRP?"

"Yes we do and it's fully equipped. The window sticker price is $56,000, but we would be happy to discuss offers and possible trade-ins."

"There are no trade-ins and this vehicle would be a company car, not for personal use, to transport our customers."

"How may we proceed? I can schedule a test drive, if you would like."

"What is your name? I will call you back in fifteen minutes."

"Ralph and what is yours?"

"Karl."

Jon next went back online and found an auto detailing service, fairly close to the dealership, that tints windows. He called them, discussed his requirements, obtained timing and cost, then made an appointment stating the car would be dropped off at their location later this afternoon. He gave his name as only Karl.

With Ralph back on the line, he made an offer, "Would you accept $52,000 cash, including taxes and license fees, for the CTS we discussed?"

"Let me check and call you back."

"I'll stay on hold."

"OK, I'll be back." Jon thought it was fortunate there was no irritating music on hold.

Ralph returned in less than five minutes, "It appears you are not a negotiator. Your offer is accepted. When can we meet to close the sale?"

"Give me your Fax number; I'll send all the registration details and call back in two hours to verify the documents and car are ready. I can be there with a cashier's check about twenty minutes after that."

Jon wrote down the number on Randy's memo and hung up, saying, "Soon we'll be able to move around this city on our own."

Silence ensued, then Lucinda offered, "Ricardo and I are hungry. He tells me there's a deli across the street. Can we bring back something?"

Katie responded, "Great idea, I'll have roast beef on sourdough with lots of mayo and onions, plus a Coke, Jon?"

"Same for me. You might check with the others if you can carry that much." He pulled a $50 dollar bill from his pocket and slid it across the table.

Don and Charlie begged off lunch, stating they didn't want to be late for a DIA meeting. They said they were most pleased with the employment contract. Randy said he brought his own lunch. Karl sat down and Jon explained about the car and made his request of Karl. Katie asked Karl where she could find a wireless printer to send the needed Fax. He showed her.

"Karl you might want to give our business bank a head's up so the check will be ready. I just emailed you the addresses of the dealership and detail shop."

The control group briefing was scheduled for 11 AM back at Justice. They took the elevator down to their building's underground parking area and there sat a gleaming Cadillac CTS.

"Looks like a Mafia staff car," Katie observed.

They felt comfortable and inconspicuous as they drove to another underground parking area at Justice.

Everyone was already seated when they walked in, including AG Gates and Security Advisor Prescott. Jon immediately went and gave General Gus a long, firm handshake. Katie did the same thing with Shelia.

Gus opened with, "Quite an adventure, if I can use that word. You seem to be relaxed and dodging all the media bullets." Katie chuckled.

"What are you willing to disclose?"

Jon took the lead, "Trust is a marvelous thing, but no recorder or notes please. I'll give you a complete run down, from planning through execution, and the aftermath. I'm sure you remember the Guzman, etc. phone calls from our previous meeting, and our month's long efforts to keep him in check using other methods. Our group decided: no more, enough is enough, we want to live our lives free from the constant fear and vigilance."

For the next fifteen minutes, Jon took them through it, not mincing any words or leaving out any significant detail. When he was finished he looked at Katie and asked, "Would you like to add anything?"

"Only to repeat our sincere thanks."

Preston Gates was shaking his head and repeating several times 'almost unbelievable'.

For the first time ever there were no questions.

Gus asked, "So where do we go from here?"

Katie said, "Very clear to me. We have a contract to provide services to this group and we're ready to fulfill our end of the deal. Our facility is complete and ready for operations. We are having an open house for all of you this coming Saturday, if that could work. I wouldn't want any of you to party on government time."

Gates was shaking his head again, this time chuckling.

"The doors open at 3 PM. The cocktail hour begins promptly at five and a buffet dinner will be served at six. If you have any target or specific geography, we would love to demonstrate."

On their way out, Katie pulled Preston and Clint aside.

"I need some advice or a recommendation concerning our open house. I'm certain you people hold many important meetings with equally important attendees that are catered. Do you know if there are any trustworthy caterers that do not ever identify the people or repeat what they might have overheard? If not you'll just have to put up with my cooking and Jon's bartending."

Clint said, "Let Preston and I think about it and we'll get back to you shortly. We appreciate your recognizing the need for confidentially when a group such as ours meets, or is seen outside."

Upon their return they found Randy and Lucinda setting up and organizing a filing system. Ricardo was seated at a room center table with a sheaf of paper in front of him. Karl was in his office on the phone with Marie.

Randy looked up at Jon, "We organized all the cell phone chatter across northern Mexico and Ricardo is reviewing the translations. Many calls and you can certainly guess the main topics, but there doesn't appear to be anything worrisome. Have a look if you want."

"I will." He sat in a chair next to Ricardo.

Katie departed with, "I'm going to write my next column and make some open house lists."

At the end of the day, all but Katie were gathered in the conference room. Karl asked, "What's up with Katie? She's bent over with her fingers flying. She looks like a woman possessed."

Jon snickered, "Then you've never seen her compose one of her columns. I've been wondering what her next topic or issue will be. I'll admit to being somewhat nervous. That's been an area of our lives where I stay out. About the only person she looks to for edits is Edgar Greeley."

Katie's furious typing was interrupted by her cell phone. She didn't recognize the number, but answered the call. It was Clint Prescott.

"Hi there and thanks for calling me."

"Are you sitting down? You sure have some powerful admirers. As you might have expected, I briefed the President after our meeting, leaving out much of the detail Jon provided. He made no comments for the record but was visibly pleased. I took the liberty, after a discussion with Preston, and received approval to use the White House staff for your open house. There is no state dinner planned any time soon and you can expect a call from Philip Larose in the morning, the head chef. The President sends his best regards."

"Oh my God, thank you."

"See you Saturday."

Katie joined the others and was barely able to control herself, as she explained what had just happened.

Randy said goodbye for the day. They decided to order pizza to be delivered. Karl and Jon cracked a beer and Katie returned to her composing. An hour later she had a rough draft. She sat back and heard the elevator door open. Karl was returning from the lobby with pizza and salads.

Katie followed him into the conference room and opened a beer for herself. To Lucinda and Ricardo she said, "Help yourselves. You're close enough to 21 and won't be carded here."

They did. It was almost 8 PM when all the food had been consumed. Katie offered to clean up and passed out her lists for the open house, requesting any feedback.

Katie was up brewing coffee in the kitchenette of their small apartment, when her phone rang – another unknown number. She answered and the caller announced he was Philip Larose from the White House staff, with a slight accent.

"I'm pleased to make your acquaintance and thanks for the call."

"How may we assist you, Ms. Ruiz?"

"It's Katie, and if you have the time now, I'll explain what I have in mind."

"Please tell me."

Katie explained the location, its facilities, the timing, and expected number of guests. She went on, "We will use our reception area as a bar and fully stock it including wine. We request the services of a bartender and appropriate glassware if possible. If not we'll go purchase it. Regarding dinner we suggest a buffet style and would never think to insult someone like you with any menu requirements. Again, we have no dishes or cutlery. Our large conference table should work well. That's about it."

"Thank you Katie for such a concise explanation. I will call you back before the end of today."

She poured some coffee and looked around wondering, *'where's my desk?'* With a second cup in hand, and still in her bathrobe, she returned to her computer to begin editing her rough draft. A short while later she looked up to find Lucinda and Ricardo at her door.

"Good morning. What's up?"

"May we interrupt?"

"Come in and sit down." Katie could almost guess what was on their minds. She had been expecting this visit. She waited for the twins to start. Lucinda held out to Katie a copy of Randy's staffing memo, now with yellow highlight marks.

Katie took it, "If you've wondered why I haven't broached this subject with you before, it's not because I'm unaware. It's because you are adults and in control of your own futures. The decision on any path forward must come from you without undue influence."

Ricardo spoke up, "We want to remain a part of your group, or should I say business now."

Lucinda added, "We are capable of these yellow highlighted jobs and hope we have shown how fast we learn."

Katie excused herself and returned with Jon who leaned against the open door jam.

"Katie and I have discussed many times how we might assist you in the future as you continue to grow and change. She told me what you just said. I have some very direct questions. Is that OK?"

"Yes" and "Yes."

"Can you see yourselves living in the DC area for the foreseeable future?"

"Yes" and "Yes."

"Are you willing to continue your formal education?"

There was a slight hesitation, then "Yes" and "Yes."

"As we have a not insurmountable issue with security clearances, would you be willing to apply for, study for and obtain American citizenship?"

"Yes" and "Yes."

Jon walked over, smiling broadly, and engaged them in a group hug, "Sign them up Katie."

Katie's face beamed, "Let's meet with Randy right after lunch. I'll get your employment contracts prepared. Please give me your visas."

In the middle of the afternoon, with Karl finishing up a few loose ends, before his trip to the liquor store with Katie's list, Philip Larose called back.

"Katie, I believe you will be satisfied with what I'm about to say. Your requested bartender has been selected and asked me to convey he is an admirer and looking forward to serving you and your guests. A wide variety of glasses and stemware will be provided. However, please allow us to provide the wine. As an accompaniment to certain foods, its selection is very important. If you provide your email address, I would be pleased to send my recommended dinner menu. We would suggest arriving about 1 PM to set up, and we will stay unseen until any serving commences. You might note on the menu there are appetizers for before and during the serving of cocktails."

"Wow!" was Katie's first reaction. She thought she heard Philip chuckle. She provided her email address and said, "I can't thank you enough, see you Saturday."

Chapter 43

On Thursday morning Jon and Katie took Lucinda and Ricardo to the local DMV office to take their written driving tests. Jon took a calculated risk and accompanied them into the building. Katie remained in the car. That afternoon they drove to the Georgetown University campus and its admissions office. Jon also accompanied them to a hastily prearranged meeting through past colleagues at the University of Washington. Jon identified himself and was greeted effusively.

Lucinda and Ricardo received all the needed admission forms, and Jon informed the registrar they would be receiving a letter of recommendation for the twins from AG Preston Gates, one of their most impressive alumni. That statement was almost overkill, but certainly had its intended effect.

On Friday, with everything in place, Jon and Katie took the car, crossed the Potomac into Virginia, and headed for the coast for a scenic drive, walk on the beach, and a long romantic lunch.

Saturday finally arrived. Randy had arrived early followed shortly thereafter by Don and Charlie. They had been invited. Their first official employment day was the following Monday. Karl was fussing around in the GEOWest room; Jon in the GEOEast room. The twins were vacuuming and cleaning. Katie didn't care that everyone saw her in her bathrobe and pillow hair.

Katie scanned the latest media stories and found nothing new or interesting; however, the speculation and conjecture continued. She hated being a 'Media Darling'. With a devious look in her eyes, she thought about her next column: *'Just you wait Henry Higgins, just you wait'*.

By 11 AM she was showered, dressed, and back at her computer putting finishing touches on the column before emailing to Edgar for comments. They were always distributed on Sundays for Monday publication. The number of syndicated outlets had grown to over 300, and her readership was estimated at 25 million. She was the seventh most read journalist in the United States.

Promptly at 1 PM her cell rang and it was Philip Larose, stating he was in the lobby needing access to her floor. She told him she would be right there. Four trips later, the people, boxes, and carts were all in the office. She was introduced to Roger who would be their bartender and four others who would be the servers. She showed them where everything was located and stepped out of their way – practiced precision took over.

An hour later, Philip Larose found Katie in front of her computer, knocked on the open door frame, and said, "Please let me show you and ask for your approval." Katie decided, despite the accent, his words were not the least bit condescending. She followed him and was beyond impressed, if there's such a thing.

In the bar area was a wine rack with displayed bottles and labels of description included. That rack was replicated in the conference room where the food would be served. The table had been set with various sizes of plates, cutlery and napkins. Ten chafing dishes were aligned along the remaining length of the table, some with burners underneath, some with ice trays. Six carts were against the window wall.

"Would you like to see their contents?" Phillip asked.

"Not necessary; I saw your emailed menu and can hardly wait to indulge my pallet."

Larose laughed, "Do you have a table, preferably circular for the reception area. We have prepared a guest arrival delight."

Katie found Ricardo and from the storage room they brought out the requested table. In a flash it was covered with white linen and silver serving trays containing a vast array of colorful, bite size somethings.

"Help yourself." Larose offered.

They did, and the melt in their mouths, not too sweet treat was enjoyed. Chocolate was obvious but not predominant. The other blend of flavors was far too complex to identify.

"Ricardo, don't you dare tell the others about these. Let's move the table into my apartment without being seen."

Both Larose and Ricardo laughed, but not before taking another treat.

Phillip said, "We are ready and will take our leave now. We have the code to your floor. Roger will return a few minutes before four and we will return at five."

Lucinda, looking absolutely dazzling, she was a natural beauty by any standard, was in the lobby to greet and control the elevator. Katie was seated on a reception area upholstered bench to be the greeter. The rest hung back in the offices.

The first two to arrive, several minutes early, were Margaret, the Justice contracts' lawyer, and Gloria from the DIA. No sooner had Lucinda sent them to the eighth floor than Shelia arrived.

Katie greeted them warmly and said, "Glad to see you three. I need some balance with all the testosterone around here. Maybe I should tell Lucinda to lock the elevator and come on up. Please enjoy a special treat from our friends at the White House." She waved towards the table.

They all had a good laugh and helped themselves.

Preston, Clint and Gus arrived a few minutes later and were sent up. Lucinda waited another ten minutes before joining the last guests on the list in the elevator: FBI James and CIA Jim.

After Katie's last greeting, the assembled throng of eight invited guests was greeted by Jon and Karl. Randy, Don and Charlie hung back. Katie thought it had been orchestrated perfectly, and Phillip's treats were a total hit.

Jon led them out of the reception area, past the conference room and offices, to the GEOWest room. Karl gave a detailed briefing. Next, they viewed the duplicate GEOEast room, and finally the analysis and data storage room.

Being the most technical of the group CIA Jim was awestruck and said so.

Jon asked, "We would like to demonstrate functionality. Is there any target that might interest you?"

CIA Jim turned to Security Advisor Prescott and in almost a whisper, asked, "Am I allowed to disclose?"

"Yes, go ahead and I have your back."

All the others stood in rapt attention.

Jim began, "We have a current and ongoing situation on the Iran, Iraq border. It seems some of the former Saddam Hussein Republican Guard soldiers, being of Shiite persuasion and closely aligned with Iran, have joined forces and are mounting attacks on our troops and other allied fighters. We don't know who's behind this or directing these battles."

Jon said, "Let's move to the GEOEast room and give me some geographical names in the area."

They did, and Jim provided three names, unsure of the pronunciation.

Jon was into Google Earth and said, "No matter, I found them. Randy or Ricardo, I just sent to the printer three GPS coordinates. Would you please focus our spot beams appropriately?"

Randy was seated in front of the controller and frequency tuner when Ricardo handed him a printed sheet.

"Our satellite is now responding, please be patient for a couple of minutes."

The awestruck group could almost sense the GEOEast satellite rotating into position.

"Locked on and receiving now," Randy announced.

Katie spoke up, "It's almost five, and should we capture anything useful, we'll call you back. Please relax and enjoy our offerings." She bade them to follow her.

Reluctantly they did and were greeted by bartender Roger, and new tables of hors d'oeuvres. The drinks and nibbles definitely relaxed the group over the ensuing forty-five minutes. Ricardo came out and beckoned Jon to follow him. They entered the GEOEast room and saw Lucinda reading translation screen output.

Randy said, "I think dinner might be delayed."

Jon looked over Lucinda's shoulder and read.

He looked at Randy and asked, "Are these instructions continuing?"

"Yes, in near real time. I am having some difficulty separating all the calls between local battlefield communications and those coming from Tehran. It's fast and furious right now."

"Ricardo, get our guests back here right now, thanks."

The room was crowded and Jon signaled for silence.

"All of you have a choice to make right now, Jim, you in particular. This system has intercepted, decoded, and translated many communications in your areas of interest. They define an imminent attack on US and allied forces from combined Iranian and dissident Iraqi forces, all directed from Tehran.

"Your choices are these, one: take our intel per the contract, have it analyzed, conduct several planning meetings, then try to obtain a decision. Realistically, this battle will be long over by then, with many US casualties. Your other choice is to believe in us and this system. I will define the intel Randy collected and Lucinda and Ricardo helped translate, and you can act upon it NOW. Understand your choices? You have ten minutes to decide."

Jon left them stunned and went to get a margarita topper. Katie squeezed his arm as he brushed by. When he returned, both Clint and CIA Jim were on their phones.

Looking at his watch, Jon said, "Your ten minutes are up. What's your decision?"

Both disconnected and asked for the summary intel.

"There is a large force of combined Iranian and Iraqi forces preparing to move out from Piranshahr, just across the northwestern Iran – Iraq border. Their orders come from a point north of Tehran. We have those coordinates and numbers.

"There is some kind of base camp of US and maybe allied forces several miles east of Erbil, Iraq. They are currently unaware or asleep at the switch. The attack is imminent. That's what we know."

Clint looked at CIA Jim and said, "Do it."

General Gus interrupted, "No Jim, just inform them. I will do it. Can I have a private room?"

Katie led him to the office she and Jon shared and closed the door behind him, then went back and coaxed the others to the conference room for some terrific entrees with matched wine.

Gloria, from DIA, seemed the most relaxed and the others took heart from her words, "Don't ever underestimate General Shepherd and his buddy, the Chairman. They can move continents in a New York minute. I almost feel sorry for those enemy soldiers, with the firepower coming their way."

Gus returned to the GEOEast room and said, "It's done. I'll have real-time reports all night. I'm hungry. Let's go see what Katie has for us."

The talents of Philip Larose and staff exceeded all expectations. The flavor and variety of the dishes defied description. There was seafood; there was beef; beyond that, unknown. The serving staff courteously removed wine glasses depending on the course, and replaced with what they considered best matched. They dined for almost two hours.

They were in the midst of saying good night when Gus' phone rang. He moved down the hallway to privacy, and everyone went on alert. Just over ten minutes later he returned.

While addressing the group as a whole, his eyes were directed at Jon and Randy, standing side by side.

"I'll make this short, with a full follow up in the morning. We have engaged the enemy with air power. They were exactly as your intel provided. The base at Erbil is on full alert and taking defensive precautions. They are also being deployed to clean up if there's anything left. If there was ever a doubt in my mind about your capability, it's gone. I'm a believer, and so should be the Erbil soldiers' families back home. Thank you Katie and group for a fabulous evening."

Lucinda came out to escort the guests, by groups, to the lobby and security locked entry door.

Both Katie and Jon awakened Sunday morning with slightly thick heads – not exactly a hangover, but close.

She rolled over, "Let's snuggle for a while. It's Sunday morning coming down. What an outstanding night."

Jon was finally forced to get up and use the bathroom. Katie looked at the bedside clock. It reported almost 10 AM and she reluctantly got out of bed and grabbed her faithful robe.

The entire office was quiet. They wondered the whereabouts of Karl and the twins.

Coffee was brewed, and instead of sitting on the deck overlooking Tucson, they chose the reception area of their office. Reluctantly Katie stood up. "I better get moving. I've got a column to file."

She got another cup of coffee and went to her office. After booting both her phone and laptop, she perused the news headlines. The complete destruction of the combined Iran, Iraq forces was the headline news. She chose not to open the articles, preferring the real story from Gus. From the lack of other headlines, she was cautiously optimistic the previous speculation about she and Jon had moved to either back printed or Web pages.

Her two emails of importance were from Gus. She could see Jon had been copied, and another from Edgar with a subject line of 'Really?' She decided to read the Gus email with him or after Jon had read it. She opened Edgar's email and read the short message:

> *'While I do understand your reasons for writing this, I'm not sure it's advisable. Give me a call after 10, my time, and I'll give you my reasons why.'*

She looked at the time on her phone and noted she could call Edgar in just over an hour. She anticipated what Edgar would say and promised herself to listen and consider. She had until 5 PM, Tucson time, to file the column but there was no alternative version available. On the flip side, one of her genetic strengths or defects is determination and particularly when instincts tell her it is right. She had no intention to discuss this with Jon. The chips will fall where they may and he promised to love me forever.

Jon, with a cup of coffee in hand, stuck his head in her open door and said, "Good morning beautiful, what are you up to?"

"Good morning to you. I was just perusing the news and biding my time to call Edgar. There's an email to us from Gus. I waited for you to read it. I will now. What they did is all over the news."

General Gus' summary was fairly short, and provided little detail. While they were dining last evening, a squadron of A10 Warthogs was scrambled from a base north of Mosul and given the Piranshahr GPS coordinates Randy and Jon provided. The Iran, Iraq fighters were on the move from this location, and the A10's engaged. The size of the force was estimated to be about 600. They were eliminated along with most of their equipment. A company of Army soldiers is currently at the site, mopping up. No further combatants have been found. Knowing that GEOEast was continuing to monitor and receive and Randy had volunteered to spend his Sunday at their facility, the General's email ended with a request to forward any new intel ASAP.

Jon said, "Just spoke with Randy and he hasn't seen many calls, but asked if Lucinda and Ricardo could help with the translations. Do you have any idea where they and Karl are? I tried his cell and it went to voicemail."

"No I don't. Sorry. I wondered the same."

Jon turned and just then his phone rang, "It's Karl. I'll take it."

Katie looked at the time and called Edgar Greeley. He picked up on the second ring.

"Morning, Katie."

"That it is. Fire away Edgar."

"I have three comments to make and then let's discuss them individually. My first reaction to reading the piece was your anger and resentments; I read it a second time trying to extract the real message and couldn't find one, only the reasons for the disappointments. I have a problem with the next to last paragraph, as it fuels the current speculation of your's and Jon's activities. Has he read this?"

"No, he hasn't and I've never used him as an editor before or even given a sneak peek. My only real anger is the complete failure of the efforts the President promised.

Resentment of the intrusions into my personal life, absolutely. I'm still digesting your other comments."

"Please, tell me your intended purpose for the column, beyond venting."

"I need to vent. I need to get it all out before I either explode or become very depressed. I can't do that with Jon and our group because we must keep holding it together for our protection and sanity. I have no other outlet. Can you understand that?"

"Yes Katie. I do. Let me make some red line edits and I highly recommend you let Jon read this before release. I'm always here for you."

"I know that and love you Edgar. Thanks."

Katie sat back and thought about the conversation for a few minutes and decided Edgar was right, this time. She walked down the hall and found them in the GEOEast room. Karl and the twins had returned. She asked where they had been.

Karl answered, "We took a drive to Chesapeake Beach, had a great breakfast and got our feet wet. I turned my phone off in the restaurant and forgot to turn it back on until we were almost back here, sorry, but it was fun."

"I was worried, but all's well. Jon, are you tied up here or can we talk for a few minutes?"

He broke away from looking over Randy's shoulder and said, "Sure, now's fine."

He followed her back to the office. Before he could sit down, Katie hugged him strongly and buried her face in his chest. Jon's first reaction was to hug her back and then wondered what was going on. Katie let go and sat down. Jon did the same. She handed him two sheets of paper, "Please read. I need your feedback."

With some trepidation Jon began to read. He read it a second time. When finished, he put the pages on the desk and said, "Man, do we keep stuff bottled up inside, and by we, I mean all of us. What do you want me to say? – besides, I love you."

Tears came to Katie's eyes.

Jon came to her, moved her hair out of the way and kissed her tears.

With the emotion drained, she told Jon about her conversation with Edgar and his upcoming edits.

"Edgar has some points so let's see his proposed changes. I wouldn't change a word. The person who wrote this piece is exactly the person I married and will love forever, unconditionally."

"Thank you, from the bottom of my heart."

Chapter 44

At 5 PM the Arizona Star released the following column to 312 syndicated outlets:

Desert Musings

June 24, 2007
by Kate Ruiz-Reynolds

This week's column departs from its usual focus on issues Americans face each day. Reluctantly this column is about me. To begin, let me attempt to connect these dots: celebrity, privacy, and the media. The common definition of a 'Celebrity' is a famous person. Famous for what, that is a key question. There are no Constitutional guarantees addressing the right to privacy. In the context of the First Amendment and the 'Freedom of the Press' principle, there are no effective privacy laws whatsoever.

I have become a Celebrity. Some celebrities seek fame and indulge themselves in their status. I do not and resent my current status. How I became a Celebrity is well known, not to be repeated here. As a member of the profession, I have a total understanding and respect for the media and its tenets. Examples of those include accuracy, objectivity, protection of sources, and respect for individuals and institutions, which brings me to the issue of privacy. What I published last year was certainly newsworthy and for the most part accurate. The release of certain biographical information about the authors is to be expected. Delving into or becoming fixated on their personal lives is not relevant and shows a complete lack of respect. There is hardly a day that goes by without some media release that invades my privacy and has placed me, my friends and loved ones in personal danger.

I began my journalism career full of purpose and idealism. I will right wrongs. I will expose crime and corruption. I will increase people's awareness and understanding of those issues that affect their lives. I will help individuals achieve their goals to the best of my ability. Today I feel like a failure.

Last August, in the Rose Garden, I was on top of the world standing beside my soul mate and friends. I believed we had accomplished something important and not simply for the benefit of governments. Please allow me to repeat what I said then.

Standing here today gives me the opportunity to reach out to those unable to cope with today's world, or who have been forgotten and left to suffer in their own misery and addictions. Coming from the background I do, let me tell you the path to self-esteem and happiness in this life is not up your nose or at the end of a pipe or needle. There is another way to feel good about yourself and cope with the difficulties of life. That way is to reach out and grab ahold of the opportunities that abound in this country. Can't see them? They're blocked from you or unreachable? Then we as a government and society have failed. The ultimate solution to our nationwide drug problem is to eliminate the demand.

The disruption we caused in drug trafficking was temporary. Today the flow of drugs into this country has nearly returned to previous levels. The demand for these drugs has not abated in any measurable way. Drug related crimes continue each day. Government has accomplished next to nothing, other than to collect and pocket an enormous sum of money. I consider my efforts to be a failure.

The Cartels have regrouped and reorganized. I have personal experience supporting this. The entire group of us became targets of these Cartels. Let me remind you of the murders of Seattle police officers Thomas Murphy and Don Minelli, and the attempted murder of Sgt. Ortiz. My husband's cabin was attacked, the Arizona Star building was attacked causing injuries and death, my home was attacked. It doesn't end there. For over ten months we have been engaged in nearly constant vigilance to protect our lives and have taken the battle to the enemy.

I will be taking a sabbatical from this column to regroup and rethink my future. Be assured sliding into apathy is not an option. Thank you to all my readers.

Monday found the office with eight occupants busy and focused on their respective responsibilities. Katie was choosing not to view the news of the day. She was organizing the employee files, including for Don and Charlie down the hall. Karl was packing up his office preparing to take a first class, non-stop commercial flight home that afternoon. Katie and Jon were indecisive about their next port of call, but determined to manage their new company and its systems for its intended purpose: continuation of and winning the war.

Katie's phone rang and she recognized the number to be AG Preston Gates.

"Good morning Katie. Have you seen the news?"

"No. what have I missed?" She clicked the *Washington Post* link and saw the front page headline.

El Chapo and El Mayo Have Escaped

"Oh my God, what happened? I haven't read the report yet."

Gates answered, "They found a tunnel from the shower room leading under the detention facility wall. Total incompetence I should add. I'm really sorry. I read your column this morning and I'm sorry about that and this government."

"I appreciate that Preston."

"All my best regards and wishes for you and Jon."

Katie did a screen print of the headline and showed it to the others. The reactions were similar: initial shock, disgust and wondering what the future held.

Jon took Katie in his arms and loud enough for everyone to hear said, “As long as we are together we will survive anything that may come our way.”

About the Author

Jeffrey Warren is a retired Professional Engineer and longtime entrepreneur having founded, operated and sold four businesses from aircraft design and manufacturing, furniture manufacturing, software architecture with database applications, to an RV park.

He is the author of several technical, non-fiction papers and books. He began writing fiction for a family audience during an interlude between companies. His first published work was an illustrated collection of musings and short stories entitled, *Stages of Life*. This novel is the sequel to *Justifiable Homicide,* published in 2017. In 2018 he published *Fifteen, Back When* and its sequel in 2019, *The Adventures of Brett and Molly*. These novels are coming of age adventures.

An avid sailor, he lives with his wife, seven horses, a donkey, and rescue dog *Scout*, on a mountain ranch in Idaho. He has followed a lifelong belief that wherever the wind may blow, you will find opportunity and gain valuable experience.

Visit at www.jwauthor.com.

About the Author

Made in the USA
Monee, IL
04 November 2026

67380507R00193